HOLLYWOOD MARINES

Celebrities Who Served in the Corps

MSgt Andy Bufalo USMC (Ret)

ISBN 978-0-9817007-6-2

First Printing – March 2010
Printed in the United States of America

www.AllAmericanBooks.com

OTHER BOOKS BY ANDY BUFALO

SWIFT, SILENT & SURROUNDED
Sea Stories and Politically Incorrect Common Sense

THE OLDER WE GET, THE BETTER WE WERE
MORE Sea Stories and Politically Incorrect Common Sense
Book II

NOT AS LEAN, NOT AS MEAN, STILL A MARINE!
Even MORE Sea Stories and Politically Incorrect Common Sense
Book III

EVERY DAY IS A HOLIDAY...
Every Meal is a Feast!
Yet Another Book of Sea Stories and Politically Incorrect Common Sense
Book IV

THE ONLY EASY DAY WAS YESTERDAY
Marines Fighting the War on Terrorism

HARD CORPS
The Legends of the Marine Corps

AMBASSADORS IN BLUE
In Every Clime and Place
Marine Security Guards Protecting Our Embassies Around the World

THE LORE OF THE CORPS
Quotations By, For & About Marines

"It is for me a touchstone of the Marine Corps' fatal glamour that there is no *ex*-Marine of my acquaintance, regardless of what direction he may have taken spiritually and politically after those callow gung-ho days, who does not view the training as a crucible out of which he emerged in some way more resilient, simply braver and better for the wear." – *Former Marine and Pulitzer Prize-winning author William Styron*

This book is dedicated to every Marine who served our Country and Corps with honor, and then returned home to raise families and make innumerable and often anonymous contributions to the fabric of this great Nation. You are the true superstars.

PREFACE

Hollywood Marines was born from what many people consider to be an annoying habit of mine. Every time a celebrity who once served in the Marine Corps appears on my television screen I feel compelled to point that fact out to all present - just in case they didn't know. I can't help myself! I knew that a couple of books had been written on the subject, and thought they would be a handy reference to keep by the TV to help flesh out the backgrounds of those famous Devil Dogs... but they proved hard to find, and the one I did locate was a bit unwieldy. The end result is *Hollywood Marines.*

The term Hollywood Marines, which was in my opinion the perfect title choice, stems from the longtime rivalry between East and West Coast Leathernecks. Those of us who attended boot camp aboard MCRD Parris Island take great delight is calling graduates of San Diego - the *other* MCRD - "Hollywood Marines," since rumor has it they are issued sunglasses and tanning oil when they enlist. Naturally the boys from 'Dago' don't take that lying down, and respond by pointing out that all Women Marines go through PI (which they do)... and further suggesting that we were *all* there as members of WM Battalion. That is not meant to be a reflection on the ability or dedication of our female counterparts, and I apologize if it offends any of my WM friends. It's just a guy thing!

We have all been touched in one way or another by famous former Marines, although most likely we didn't know it at the time. I saw Buddy Rich play his drums on the stage at my high school when I was a teenager, and even interviewed Don Imus for the school paper while he was on the air with WABC in New York City - but in neither case did I know they had once been Marines.

It would be virtually impossible to include everyone in the public eye who has worn the Eagle, Globe and Anchor, so I have been forced to limit the celebrity Marines in these pages to those who, in my opinion, would be best known to the general public. A good example is my selection of major league baseball players. An argument can be made that *everyone* who has played the game has a measure of notoriety, and although that is true keep in mind that while the website **baseballinwartime.com** lists more than *seventy* players who served as Marines in WWWII, only Ted Williams, Gil Hodges, Hank Bauer and Jerry Coleman would be known to most fans.

Great care has also been taken to exclude those who are often mistakenly believed to have served in the Corps, like Don Knotts (who was never a Drill Instructor) and Mister Rogers (who was never a sniper). Also left out are some, such as Dan Rather and boxer Riddick Bowe, who tried to be Marines but failed to cut the mustard. I have not so penalized others, notably Buddy Rich and Ernie Hudson, who had their service cut short due to medical problems over which they had no control.

Also conspicuously absent are Marines who have made a name for themselves in business, government and public service. Not to worry though - they will be the subject of a separate book in the near future!

Semper Fi!
Tampa, Florida
September 19th, 2009

TABLE OF CONTENTS

FILM, TELEVISION AND RADIO ...15
Don Adams ...17
Bea Arthur ...22
Donald P. Bellisario ...25
Joseph Bologna ...28
Hugh Brannum ...30
Wilford Brimley ...33
Drew Carey ...36
Macdonald Carey ...44
Barry Corbin ...47
Don Cornelius ...52
Brian Dennehy ...54
Bradford Dillman ...58
Dale Dye ...60
David Eigenberg ...63
R. Lee Ermey ...65
Mike Farrell ...69
Glenn Ford ...72
Christopher George ...78
Scott Glenn ...81
James Gregory ...84
Clu Gulager ...86
Gene Hackman ...88
Sterling Hayden ...2
Lloyd Haynes ...95
Louis Hayward ...97

George Roy Hill ...100
Ernie Hudson ...104
Don Imus ...106
Bob Keeshan ...108
Harvey Keitel ...111
Brian Keith ...114
Dan Lauria ...117
Jim Lehrer ...119
Joe Lisi ...122
Lee Marvin ...123
Tim Matheson ...128
Ed McMahon ...130
Steve McQueen ...134
Alvy Moore ...146
Warren Oates ...148
Hugh O'Brian ...151
Gerald S. O'Loughlin ...155
Pat Paulsen ...157
Sam Peckinpah ...160
George Peppard ...168
Tyrone Power ...172
Hari Rhodes ...180
Rob Riggle ...183
Pernell Roberts ...185
John Russell ...88
Robert Ryan ...190
George C. Scott ...193
Bernard Shaw ...199
Bo Svenson ... 202
Bobby Troup ... 204
James Whitmore ... 207
Larry Wilcox ... 210
Steve Wilkos ... 214
Montel Williams ... 217
Jonathan Winters ... 220
Burt Young ... 223

Hollywood Marines

SPORTS ... 225
Dusty Baker ... 227
Carmen Basilio ... 229
Hank Bauer ... 231
Patty Berg ... 236
Rod Carew ... 238
Roberto Clemente ... 241
Jerry Coleman ... 244
Eddie Collins ... 247
Art Donovan ... 252
Bill Fitch ... 255
Hayden Fry ... 257
Ernie Harwell ... 262
Elroy Hirsch ... 266
Gil Hodges ... 268
Keith Jackson ... 272
Mills Lane ... 275
Tommy Loughran ... 278
Bob Mathias ... 280
Tug McGraw ... 283
Billy Mills ... 288
Rick Monday ... 291
Jim Mora ... 295
Ken Norton ... 299
Bum Phillips ... 303
Barney Ross ... 305
Tom Seaver ... 311
Leon Spinks ... 316
Richard Steele ... 320
Lee Trevino ... 322
Gene Tunney ... 325
Bill Veeck ... 328
Mike Weaver ... 333
Chuck Wepner ... 336
Jo Jo White ... 339
Ted Williams ... 342

***THE ARTS* ... 347**
James Brady ... 349
Art Buchwald ... 352
Orville Burell (aka "Shaggy") ... 355
Richard Diebenkorn ... 357
Andre Dubus ... 359
Don and Phil Everly ... 363
Freddie Fender ... 367
Bill Gallo ... 371
Josh Gracin ... 373
George Jones ... 375
Robert Kiyosaki ... 377
Robert Ludlum ... 380
Buddy Rich ... 384
Mark Shields ... 387
Leon Uris ... 389
Joseph Wambaugh ...392

***PHOTO GALLERY* ... 397**

FILM, TELEVISION & RADIO

DON ADAMS
Would You Believe...?

Don Adams (April 13, 1923 - Sept 25, 2005), born Donald James Yarmy, was an actor, comedian, game show panelist and occasional director who, in his five decades of television, was best known for his role as Maxwell Smart (Agent 86) in the TV situation comedy *Get Smart* (1965-1970, 1995), for which he also directed and wrote. Adams won three consecutive Emmy Awards for his portrayal of Smart (1967-1969). He also provided the voices for the animated series *Tennessee Tuxedo* (1963-1966) and *Inspector Gadget* (1983-1985) as their respective title characters.

Adams was born in New York City, the son of Consuelo (née Dieter), a housewife, and William Yarmy, a soda clerk. His father was a Hungarian Jew and his mother a Roman Catholic of Dutch and Irish heritage. He was not raised in any religion, but after praying to God for survival during his blackwater fever illness he went from being an atheist to a life-long Catholic.

After Don dropped out of New York City's DeWitt Clinton High School at age eighteen (his senior year) in 1941 one of his first jobs was as a theatre usher, and he later worked as a comic and mimic, taking the stage name of Adams after marrying singer Adelaide (Dell) Efantis, who performed under the name Adelaide Adams. After their divorce he still went by the stage name "Adams," explaining (perhaps comically) he kept it because he got tired of being last during alphabetical auditions.

Adams served in the Marine Corps during World War II in the Pacific Theatre. He was wounded during the Battle of Guadalcanal, became ill with blackwater fever, and nearly died. Upon his recovery and return to the States he held various jobs, and once mentioned in an interview on Canadian television that he faked prestigious college credentials and an engineering background in order to work as an engineer designing underground sewers - but his lack of training was not discovered for six months. He also served as a Marine Corps drill instructor, and was commended by his superiors for being able to exceed the performance of his recruits at every required physical task.

His work on television began in 1954, when he won *Arthur Godfrey's Talent Scouts* with a stand-up comedy act written by boyhood friend Bill Dana. In addition to appearing on numerous comedy, variety, and dramatic series, Adams had a role on the NBC sitcom *The Bill Dana Show* (1963-65) as a bumbling hotel detective named Byron Glick - a character nearly identical to what was soon to become "Maxwell Smart" on *Get Smart*.

Creators Buck Henry and Mel Brooks wrote *Get Smart* as the comedic answer to all of the successful 1960s spy television dramas such as *The Man from U.N.C.L.E., The Avengers, I Spy* and *It Takes a Thief.* They were asked to write a spoof that combined elements from two of the most popular film series at the time - the *James Bond* and *Pink Panther* (Inspector Clouseau) movies.

Get Smart had been written for Tom Poston to be piloted on CBS, but when CBS turned the show down it was picked up by NBC, which cast Adams in the role because he was already under contract but without a show. When it debuted in 1965 it was an immediate hit.

Adams endowed the character with his voice and clipped, unique speaking style. Co-star Barbara Feldon, who played the role of Smart's partner Agent 99, said, "Part of the pop fervor for Agent 86 was because Don did such an extreme portrayal of the character that it made it easy to imitate." Adams created many popular catch-phrases for Smart (some of which were in his act prior to this show) including "Sorry about that, Chief," "And loving it!" "Would you believe...?" and "Missed it by *that* much." These helped make the series a hit in over one hundred countries.

In addition to acting, Adams also produced and directed some episodes of the show. Off the set he occasionally feuded with Jay Sandrich, who served as writer. For his portrayal of Agent 86 he was nominated for Emmys four seasons in a row between 1966 and 1969 and won three for Outstanding Continued Performance by an Actor in a Leading Role in a Comedy Series.

For its final season the show moved to CBS, though ratings began to decline as spy series went out of fashion. *Get Smart* was canceled in 1970 after 138 episodes.

Adams was very happy about the show's cancellation as he wanted to move on to other projects, but his ventures after *Get Smart* were less successful, including the comedy series

The Partners (1971-1972), a self-titled game show called *Don Adams Screen Test* (1975-1976) and three attempts to revive the *Get Smart* series in the 1980s. Even his movie, *The Nude Bomb*, was a box-office failure. Adams was typecast as Maxwell Smart and unable to move beyond that image, though he did have success as the voice of *Inspector Gadget.*

He continued to make the majority of his income from his work on stage and in clubs, and because Adams had chosen co-ownership of the *Get Smart* property over an increased salary during the series' production period he received regular income in later years.

The Don Adams Screen Test was a syndicated game show which lasted twenty-six episodes during the 1975-76 season. The show was done in two fifteen-minute segments, in each of which a randomly selected audience member would 'act' to re-create a scene from a Hollywood movie, such as the bat scene from *The Lost Weekend* or the beach scene from *From Here to Eternity*, aided by Adams and other guest celebrities. Hokey effects, bad timing, forgotten lines, prop failures and the celebrity's ad libs were maximized for comic effect. At the end of the program the 'screen test' of each of the two contestants would be played, with audience reaction determining the winner - who received a trip to Hollywood and a *real* screen test.

Adams worked as a voice actor in *Tennessee Tuxedo and His Tales* (1963-1966), but he was more famous as the voice of *Inspector Gadget* in the initial run of that television series (1983-1985) and the Christmas special, as well as in later reprises. He even voiced himself in animated form for a guest shot in an episode of Hanna-Barbera's *The New Scooby-Doo Movies,* which first aired on October 13, 1973.

Adams attempted a situation-comedy comeback in Canada with *Check it Out!* in 1985. The show ran for three years in Canada, but was not successful in the United States. In an *A&E Biography*, Adams said he made more money working on the series, better than on *Get Smart*. He reprised his Maxwell Smart role on *Get Smart* for FOX in 1995, which co-starred Barbara Feldon and rising star Andy Dick as Max's & 99's only son. Unlike the original version, this show did not appeal to younger viewers and was canceled after only six episodes. He later went on to voice the character of Principal Hickley in the late-90s Disney cartoon, *Pepper Ann.*

In 2003 Adams joined a *Get Smart* tribute at the Museum of Television and Radio. Also appearing at the convention were surviving stars of *Get Smart* Barbara Feldon, Bernie Kopell and Dick Gautier.

Adams stated in interviews that his famous "clippy" voice characterization was an exaggeration of the speaking style of actor William Powell. Occasionally, he also enjoyed doing a more explicit impersonation of Ronald Coleman.

Adams was married and divorced three times: to Adelaide "Dell" Efantis (1949-1958), Dorothy Bracken (1960-1973) and Judy Luciano (1977-1990).

Don Adams died on September 25, 2005 in Los Angeles from a lung infection following a long battle with bone lymphoma, and was interred in the Hollywood Forever Cemetery in Hollywood. He was survived by six of his seven children, along with five grandchildren and three great-grandchildren.

BEA ARTHUR
Maude

Beatrice "Bea" Arthur (May 13 1922-April 25, 2009) was a comedienne, actress and singer. In an ongoing career spanning six decades Arthur was perhaps best remembered for her trademark role as title character Maude Findlay on the 1970s sitcom *Maude*, and also for playing Dorothy Zbornak on the 1980s sitcom *The Golden Girls*.

Arthur was born Bernice Frankel to Philip and Rebecca Frankel in New York City on May 13, 1922. Her family soon moved to Maryland, where her parents operated a women's clothing shop. She became a medical technologist before World War II when she volunteered for the Marine Corps, becoming one of its first female recruits. She then attended the now-defunct Blackstone College in Blackstone, Virginia where she was active in drama productions.

Arthur began her acting career as a member of an off Broadway theater group at the *Cherry Lane Theatre* in New York City in the late 1940s. On stage, her roles included

Lucy Brown in the 1954 off-broadway premiere of Kurt Weill's *Threepenny Opera,* Yente the Matchmaker in the 1964 premiere of *Fiddler on the Roof* on Broadway, and a 1966 Tony Award-winning portrayal of Vera Charles to Angela Lansbury's *Mame.* She reprised the role in the 1974 film version opposite Lucille Ball. In 1981, she appeared in Woody Allen's *The Floating Lightbulb.*

In 1972 Arthur was cast as the title character on the television series *Maude.* She played Maude Findlay, an outspoken liberal living in the affluent community of Tuckahoe, New York with her husband, Walter (Bill Macy) and divorced daughter Carol (Adrienne Barbeau). The show was a spin-off from *All in the Family,* on which Arthur had appeared a couple of times in the same role, playing Edith Bunker's cousin - a feminist, and antithesis to the bigoted, conservative Archie Bunker (Carroll O'Connor), who described *Maude* as a "New Deal fanatic." Her role garnered several Emmy and Golden Globe nominations, including her Emmy win in 1977 for Outstanding Lead Actress in a Comedy Series.

After appearing in the short-lived 1983 sitcom *Amanda's* (an unsuccessful US version of the British hit series *Fawlty Towers*), Arthur was cast in the hit sitcom *The Golden Girls* in 1985. She played Dorothy Zbornak, a divorced substitute teacher living in a Miami house owned by Blanche Devereaux (Rue McClanahan). Her other roommates included widow Rose Nylund (Betty White) and Dorothy's Sicilian mother, Sophia Petrillo (Estelle Getty). Getty was actually a year younger than Arthur in real life, and was heavily made up to look significantly older. Dorothy had a caustic sense of humor, and was prone to making witty and sarcastic wisecracks. The series was a huge hit, remaining a top ten ratings fixture for six seasons. Her performance led to several Emmy nominations over the course of the series and a win in 1988. Arthur decided to leave the show after seven years, and in 1992 the show was moved from NBC to CBS and retooled as *The Golden Palace* in which the other three actresses reprised their roles. Arthur made a guest appearance in a two-part episode, but the show only lasted for one season before it was cancelled.

In 2002 she made a triumphant return to Broadway, starring in *Bea Arthur on Broadway: Just Between Friends*, a collection of stories and songs (with musician Billy Goldenberg) and based on her life and long career. The show was nominated for a Tony award for Best Special Theatrical Event, but lost to *Elaine Stritch At Liberty*.

Arthur was inducted into Academy of Television Arts & Sciences Hall of Fame in 2008. On June 8, 2008 *The Golden Girls* was awarded the 'Pop Culture' award at the Sixth Annual TV Land Awards. Bea accepted the award with co-stars Rue McClanahan and Betty White.

Arthur was married twice, first to author Robert Alan Arthur, whose surname she took and kept though with a modified spelling, and second to director Gene Saks from 1950-1978 with whom she adopted two sons, Matthew (an actor) and Daniel (a set designer).

Bea Arthur died of cancer in California on April 25, 2009 at the age of eighty-six.

DONALD P. BELLISARIO
Always Faithful

Donald Paul Bellisario is a television producer and screenwriter who was born on August 8, 1935 in Cokeburg, Pennsylvania to a Serbian mother and Italian father. He served in the Marine Corps from 1955 to 1959, attained the rank of Sergeant, and received the Marine Corps Good Conduct Medal.

Bellisario earned a bachelor's degree in journalism at Pennsylvania State University in 1961, and also attended Butler Community College in El Dorado, Kansas. In 2001 he was named a Distinguished Alumnus - the highest honor bestowed on a graduate of Penn State. In 2006, Bellisario endowed a one million dollar Trustee Matching Scholarship in the Penn State College of Communications. He recalled, "Growing up in a hardscrabble western Pennsylvania coal mining town, I know firsthand the sacrifices that are made to give a son or daughter a university education... and as a Marine veteran who returned to Penn State with two small children and little money, I remember all too well that struggle. It's my hope that this scholarship will also ease the financial burden of other young men and women who have defended our country to attain their academic goals."

Bellisario worked in advertising for fifteen years as a copywriter and a creative director. He then began working in Lancaster, Pennsylvania before moving to a major agency in Dallas, Texas. From there, he made what he has described as his "big gamble" - deciding to move to Hollywood to pursue screenwriting and production.

After working under such television producers as Glen A. Larson, Bellisario adopted some of their production techniques - most notably utilizing a small "pool" of actors which he uses for his many productions.

Over the years he created several successful TV series, including *Magnum, P.I., Airwolf, Quantum Leap, JAG*, and *NCIS*. Bellisario was also the writer/producer of *Black Sheep Squadron*, which was based upon the real life exploits of Marine Corps Major Gregory "Pappy" Boyington as a fighter pilot in the South Pacific during WWII.

A trademark motif which can be found throughout most of Bellisario's work is the tendency for the protagonist to be a current or former member of the United States Armed Forces. This obviously stems from his own service in the Marine Corps. Examples are Tom Selleck's character in *Magnum, P.I.,* Thomas Magnum, who is a U.S. Naval Academy graduate, former SEAL officer, and Vietnam veteran. In *Airwolf* Stringfellow Hawke is a Vietnam Veteran and is looking for his brother who is missing in action. Commander Harmon "Harm" Rabb, Jr., the main character of *JAG,* is also a Naval Academy graduate and former Naval Aviator, and NCIS's main character, Leroy Jethro Gibbs, is a retired Marine Corps Gunnery Sergeant and sniper.

Another connection all of his main characters have is August 8th birthdays. Notably, the last episode *of Quantum Leap* takes place on August 8, 1953, Sam Beckett's birthday, and the reverse of the last two digits of Bellisario's own birth year.

In an interview with Sci-Fi Channel in the late 1990s, Bellisario said he was inspired to create *Quantum Leap* in 1988 after reading a novel about time travel. His service in the Marine Corps alongside John F. Kennedy's assassin, Lee Harvey Oswald, was the basis for a two-hour (two-part in reruns) episode of *Quantum Leap.* "Lee Harvey Oswald" (the fifth season premiere) originally aired September 22, 1992.

Bellisario was married to actress and producer Deborah Pratt, who starred in several of his shows. He currently resides in Studio City, California with his fourth wife, Vivienne. He has a son, Michael Bellisario, who had a recurring role as Midshipman Michael Roberts on the military drama *JAG,* and who also played Charles 'Chip' Sterling on *NCIS*. He also has a daughter, Troian Bellisario, who played Sarah McGee on *NCIS*. Sean Murray, who plays Timothy McGee on the same series, is his stepson.

Bellisario has identified himself as socially liberal and fiscally conservative, and is "especially" conservative towards the Military. He received a star on the Hollywood Walk of Fame in 2004, which he showed off on an episode of *JAG*.

JOSEPH BOLOGNA
Blame it on Rio

Joseph Bologna is an actor who was born in Brooklyn, New York on December 30, 1934. He is married to actress Renée Taylor, with whom he has a son named Gabriel. Bologna has a long history in film and television. His breakthrough film, *Lovers and Other Strangers*, which was co-written with his wife, was based on the true-life circumstances of organizing a wedding on short notice with the involvement of his Italian extended family and Renée's Jewish clan. Several relatives even performed as extras in the final cut. Then a year later, in 1971, the couple again collaborated to write and perform in the movie *Made for Each Other*.

Bologna remained close with his old-neighborhood aunts and uncles after becoming successful, and two of them were slightly famous on their own. His Uncle Pat was 'Blacky the Bootblack,' whom Joseph Kennedy credited as his main influence when he sold all of his stock holdings in the summer of 1929 (the market crashed in October), and his aunt Pauline was one of the best-known chefs to the stars, working for Jackie Gleason, Burt Reynolds and many other luminaries.

A decade after *Lovers* Bologna's aunt Pauline chastised him for starring in the nudity-filled *Blame it on Rio,* co-starring Michael Caine. Bologna replied, "Blame it on *me*, it's the last time I invite Aunt Pauline to a film premiere." Bologna's other best-known film role is as temperamental 1950s television star Stan 'King' Kaiser in the 1982 comedy hit *My Favorite Year*, starring Peter O'Toole.

Well known as both a writer and an actor, Bologna dates his interest in the theater from his student days at Brown University when a casting notice called for "non-actor" types to fill roles in a stage production of *Stalag 17.* He landed the leading part, but did not act again for ten years. Bologna graduated from Brown with a degree in art history, and a tour in the Marine Corps followed. After being discharged from the service he then started directing short films and writing special comedy material. "A monologue is the hardest thing in the world to write, because you're only as good as your last joke," explains Bologna. "That's why comedians are so neurotic."

Some of his other film credits include roles in *Cops and Robbers, The Big Bus* and *Chapter Two.* He also co-starred with Taylor in the Emmy-winning television special *Acts of Love and Other Comedies*, which they wrote together, and then starred in the made-for-television movie *Torn Between Two Lovers* with Lee Remick before reuniting with Taylor in the critically acclaimed Broadway hit *It Had To Be You.*

HUGH BRANNUM
Mr. Greenjeans

Hugh "Lumpy" Brannum was born on January 5, 1910 in rural Sandwich, Illinois. He eventually moved to California with his family and began playing music early on, initially focusing on brass instruments but moving on to the guitar and banjo and the instrument that was to become his specialty, the upright bass. During his college years at Redlands University Brannum became interested in jazz and, after graduation, played bass in various bands on the West Coast and at a small radio station.

He is best known as Mr. Green Jeans, the farmer and animal lover who played sidekick to Captain Kangaroo. The duo, and a number of other regular "guests" including Grandfather Clock and hand puppets like Mr. Moose, were watched by several generations of children as the Captain Kangaroo television show ran from the mid-'50s through the mid-'80s, making it one of the longest-running children's TV series ever. As Mr. Green Jeans, Brannum introduced a different live animal each episode and taught children about caring for the earth while talking about farming. Before this well-known stint, Mr. Green Jeans was known simply as Hugh Brannum, the jazz musician, and Uncle Lumpy, a children's storyteller in the late '40s and early '50s.

Brannum joined the Marines during WWII, and played in a Marine band led by Bing Crosby's brother Bob. After he got out of the service he played with the *Four Squires*, who were eventually hired by bandleader Fred Waring. Even after the other *Squires* members moved on, Brannum remained on in *Fred Waring and His Pennsylvanians*, living with his wife in a small Pennsylvania town where he took up gardening in addition to his music. The Waring's group had a regular, and well-received, weekday radio gig where Brannum first met fellow Marine Bob Keeshan (the future Captain Kangaroo), who was an employee at the station. During the program Brannum also narrated stories under the pseudonym of Uncle Lumpy as a weekly short feature for children.

This weekly segment, with music supplied by himself and *Fred Waring and His Pennsylvanians*, focused on the adventures of the character Little Orley. This popular segment led to a number of 78's on the Decca and Vocalion labels in the late '40s and early '50s. These recordings were intensely loved by those who grew up with them, but were only made available again briefly in the '60s, subsequently disappearing from popular culture, forgotten to all but collectors and fans. In 1954 Brannum hosted a local TV show called *Uncle Lumpy's Cabin*, and the following year joined up with Keeshan for *Captain Kangaroo*, playing the characters Mr. Green Jeans, Mr. Bainter the Painter, and Percy.

Hugh Brannum died in East Stroudsberg, Pennsylvania on April 19, 1987. His recordings as Uncle Lumpy were finally reissued on CD over a decade later, with *Little Orley Stories* and *More Little Orley Stories* released in 2000 by a children's toy company, Uncle Goose.

WILFORD BRIMLEY
I'm Harry!

Allen Wilford Brimley is an actor who was born on September 27, 1934 in Salt Lake City, Utah where his father worked as a real estate broker. He is best known for appearing in such films as *The China Syndrome* and *Cocoon,* and is familiar to TV viewers for appearing in television commercials for Quaker Oats and Liberty Medical, a medical company which provides supplies for diabetes.

Brimley enlisted in the Marine Corps during the Korean Conflict, and spent three years in the Aleutian Islands.

A true Westerner, the young Brimley worked as a ranch hand, wrangler and blacksmith to support his family, and his first exposure to show business was shoeing horses for stables that furnished animals for movie and TV Westerns. He also worked for a time as a bodyguard for Howard Hughes. When he returned to Los Angeles after living in Idaho for a few years Brimley began working as a riding extra for Westerns during the mid-1960s and formed a lasting friendship with up-and-coming actor Robert Duvall - who urged the cowboy to pursue acting as a career. He continued to work as an extra until stunt work earned him his Screen Actors Guild card, making it possible for him to land small parts in movies like *True Grit* and *Lawman*, and eventually a recurring role on the CBS series *The Waltons*. Then, at the urging of series star Ralph Waite, he became a charter member of Waite's Los Angeles Actors Theater.

Brimley often plays a gruff or stodgy old man, notably on the 1980s drama series *Our House*. His first characterization was in *Absence of Malice*, in which he played a small but key role as curmudgeonly, outspoken Assistant U.S. Attorney General James A. Wells, He expanded on this characterization in Robert Redford's *The Natural*, as the world-weary manager of a hapless baseball team.

Brimley was cast in the 1983 film *Tender Mercies* at the urging of his good friend Robert Duvall, who was not getting along well with director Bruce Beresford and wanted "somebody down here that's on my side, somebody that I can relate to." Beresford felt Brimley was too old for the part but eventually agreed to the casting and Brimley, like Duvall, clashed with the director. During one instance when Beresford tried to advise Brimley on how Harry would behave Duvall recalled him responding, "Now look, let me tell you something, *I'm* Harry. Harry's not over *there*, and Harry's not over *here*. Until you fire me or get another actor, *I'm* Harry, and whatever I do is fine… 'cause *I'm Harry.*"

After being diagnosed with diabetes in 1979 Brimley began working to raise awareness of the disease, and the American Diabetes Association (ADA) honored him in 2008 with an award to recognize his lifetime of service in this cause. Brimley has been especially active visiting Veterans

Administration hospitals and communities to advise patients on how to manage their disease.

Wilford Brimley is an activist, paying from his own funds for ads to have Utah allow horse-race gambling, and he was actively opposed to the banning of cockfighting in New Mexico. He lent his support to John McCain in the 2008 U.S. presidential election, and in the days leading up to his selection of a running mate McCain jokingly stated that he would pick Brimley. "He's a former Marine and great guy and he's older than I am, so that might work," McCain said.

DREW CAREY
Cleveland Rocks!

Drew Allison Carey is a comedian, actor and game show host who was born on May 23, 1958. After serving in the Marine Corps and making a name for himself in stand-up comedy, Carey eventually gained popularity starring on his own sitcom, *The Drew Carey Show,* and serving as host on the U.S. version of *Whose Line Is It Anyway?,* both of which aired on ABC.

Carey has appeared in several films, television series, music videos, a made-for-television film, and a computer game. He is interested in and involved with a variety of sports and has worked as a photographer at U.S. National Team soccer games. He currently hosts the game show *The Price Is Right,* which airs on CBS.

Carey was the youngest of Lewis and Beulah Carey's three boys and was raised in the Old Brooklyn neighborhood of Cleveland, Ohio. When he was eight years old his father died from a brain tumor. According to his autobiography he had six toes on his right foot and played the cornet and trumpet in the marching band of James Ford Rhodes High School, from which he graduated in 1975.

Drew continued on to college at Kent State University and was expelled twice for poor academic performance. He left KSU after just three years, but not before becoming a member of Delta Tau Delta Fraternity. After leaving the university Carey joined the Marine Corps Reserve in 1980 and served for six years. He also relocated to Las Vegas for a few months in 1982, and for a short time worked as a bank teller and a waiter at Denny's.

In 1985 he began his comedy career by following up on a suggestion by David Lawrence (a disc jockey friend who had been paying Drew to write jokes for David's radio show in Cleveland) to go to the library and borrow some books on how to write jokes. After winning an open-mic contest in 1986 he became MC at the Cleveland Comedy Club, and for the next few years he performed at multiple comedy clubs in Cleveland and Los Angeles. He was first brought into the national eye as a comedian when he competed in the 1988 *Star Search*, which was hosted by fellow Marine Ed McMahon. Carey was working as a stand-up comedian in 1991 when he appeared on *The Tonight Show* starring Johnny Carson. His performance that night impressed Carson so much that he invited Carey to the couch next to his desk - a rare honor for any comedian. Carey claims he reached the limit on his credit card the next day returning calls from interested casting directors, and credited Carson with making his career. In that same year Carey also appeared on the *14th Annual Young Comedians Special* on HBO and made his first appearance on *Late Night with David Letterman*. Then in 1994 Carey wrote his own stand-up comedy special, which aired on Showtime entitled *Drew*

Carey: Human Cartoon, and for which he won a Cable ACE Award for Best Writing.

With the success from his early stand-up career, he subsequently appeared in a number of supporting roles on television shows and developed the character of a hapless middle-class bachelor. In 1994 Carey co-starred with John Caponera in *The Good Life*, a short-lived sitcom that aired on NBC. After the show's cancellation Carey joined up with writer Bruce Helford (who was also a writer for *The Good Life*), who gave Carey a job as a staff writer for *The Gaby Hoffman Show.*

Carey and Helford developed the storyline for *The Drew Carey Show*, which they produced together. The show premiered on September 13, 1995 on ABC. In his autobiography he revealed his frustration with having to deal with censors and being unable to employ the off-color humor common to his stand-up routines. Carey initially earned $60,000 per episode in the first seasons, then renegotiated for $250,000, and by the final season he was earning $750,000 per episode. The show had high ratings for its first few seasons, but declining ratings and production costs of around three million dollars per episode eventually resulted in its cancellation. The program had a total of 233 episodes and lasted from 1995 to 2004 for nine complete seasons. Carey was the only actor to appear in every episode.

In 1998 Carey hosted the American version of the improvisational comedy show *Whose Line Is It Anyway?* He would announce the improv guests, direct the games, and then would usually involve himself in the final game of the show. The show ran for a total of 215 episodes between 1998 and 2006. In 1998 the New York Friars' Club made Carey the newest inductee of the group's *Comedy Central Roast.* His friend Ryan Stiles (who costarred in *The Drew Carey*

Show and *Whose Line Is It Anyway?*) served as the Roastmaster. Due to Carey's income from two shows (*Whose Line Is It Anyway?* and *The Drew Carey Show*), Forbes included him on their list of highest-paid entertainers of 1998, which was 24th with $45.5 million.

Carey has also helped to create the Improv All-Stars, which is composed of eleven other members. The members of the group have joined Carey in both of his two improv shows, *Whose Line is it Anyway?* and *Drew Carey's Green Screen Show* and some had major roles or guest starred on his previous shows.

In 2007 Carey began hosting game shows, and on April 13, 2007 he was selected to host the CBS game show pilot *Power of 10,* where contestants could win a potential $10,000,000 (annuity) top prize.

After taping the pilot episode for *Power of 10*, he was contacted by CBS regarding the opening caused by the retirement of Bob Barker at *The Price Is Right.* He initially turned down the offer to host, but on July 23, 2007 Carey officially announced on the *Late Show with David Letterman* that he would succeed Barker as host of the program beginning in the fall of 2007. The game show also includes a local connection, as announcer Rich Fields is from the Cleveland area. His first episode of *The Price Is Right* was taped on August 15, and his shows began airing on Monday, October 15, 2007. In response to replacing Barker as host of the game show, Carey stated "You can't replace Bob Barker. I don't compare myself to anybody... it's only about what you're doing and supposed to do, and I feel like I'm supposed to be doing this."

On April 22, 1999 at Disney's Hollywood Studios (known as Disney-MGM Studios at the time), one of the parks that make up the Walt Disney World Resort in Florida, Carey debuted a twelve minute attraction entitled *Sounds Dangerous!* In the show a camera follows Carey through a day as an undercover detective when his video camera fails and the audience is left in complete darkness wearing earphones, following his adventure through sound cues. The attraction is presently running.

In 1999 Carey made a cameo appearance in Weird Al Yankovic's music video *It's All About the Pentium,* and in 2004 he made an appearance for Fountains of Wayne's video *Mexican Wine.*

On May 7, 2000 Carey's made-for-TV film *Geppetto* debuted for an ABC presentation of the Wonderful World of Disney. The film was an adaptation of *Pinocchio* and included actor Wayne Brady, who had joined Carey on *Whose Line is it Anyway?* and a few episodes of *The Drew Carey Show.* Carey took singing lessons to prepare for the role.

He provided the entertainment for the 2002 Annual White House Correspondents dinner. Once he completed his standup routine for the 1,800 guests President George W. Bush made a joke of his own, noting Carey's improv work, "Drew? Got any interest in the Middle East?" In 2003 he joined Jamie Kennedy to host the WB's live special *Play for a Billion,* which was sponsored by Pepsi.

Also in September 2003, Carey led a group of comedians, including Blake Clark and the *Drew Carey Show's* Kathy Kinney, on a comedy tour of Iraq. Carey was very well received, as was Clark, for their previous military service.

On June 8, 2006, *Drew Carey's Sporting Adventures* debuted on the Travel Channel. Carey travels throughout Germany to photograph multiple FIFA World cup soccer games while he immerses himself in the culture of towns and states he visits.

Carey has a history of writing throughout his career including developing his stand-up comedy routines and then moving on to assist in writing sitcoms. In 1997 Drew published his autobiography, *Dirty Jokes and Beer: Stories of the Unrefined* where he shared memories of his early childhood and of his father's death when he was eight. He also revealed that he was once molested, had suffered bouts of depression, and had made two suicide attempts by swallowing a large amount of sleeping pills. He also wrote of his college fraternity years while attending Kent State University, and of his professional career up to that time. The book featured large amounts of profanity and, as the title suggests, includes multiple dirty jokes (there was one at the start of each chapter) and references to beer. The book was featured on the New York Times bestseller list for three months.

As a former Marine reservist, he adopted his crew cut hair style during his time in the service. Carey has had refractive surgery to correct his vision and therefore did not really require glasses (any glasses he wore in public were merely props to help the audience recognize him). However, while this was true for several years, on the May 17, 2006 episode of *Jimmy Kimmel Live* he revealed that when he turned forty he actually developed a need for bifocals.

Politically, Carey is known for his Republican leanings and has also expressed support for the Libertarian Party. *The Drew Carey Show* often presented a libertarian critique of political correctness, government regulations, racism, sexism, and homophobia, with storylines involving Carey's cross-dressing brother and dating a bisexual woman.

Since the show ended its nine-year run on ABC in 2004 Carcy has clarified that he is more of a conservative with libertarian leanings, and that he presented himself as a libertarian to avoid what some conservative critics of the entertainment industry claim is a general Hollywood bias against conservatives. On the August 18, 2006 *Penn Radio* show with Penn Jillette, Carey did however say he was indeed libertarian. He has expressed distaste for the Bush administration's running of the War in Iraq, specifically on the September 15, 2007 episode of *Real Time with Bill Maher*, and supported Ron Paul in the 2008 Republican primary.

Carey is known for being a devoted Cleveland Browns, Indians and U.S. Soccer fan, and in 1999 he was part of the pregame ceremonies at the first game of the new expansion Cleveland Browns televised on ESPN. When he promoted *The Drew Carey Show* in 1995, at the same time the Indians were making a miraculous run at the World Series, he poked fun at the rest of baseball by saying, "Finally, it's *your* team that sucks!" He also showed his support for the team by throwing out the first pitch at an August 12, 2006 Indians game against the Royals. He was rewarded by the Cleveland Indians for being "the greatest Indians fan alive" with a personal bobblehead doll made in his likeness and given to fans. Carey responded to his bobblehead likeness by saying "Bobblehead Day, for me, is as big as getting a star in the Hollywood Walk of Fame."

In 2001 Carey was the first TV star (as opposed to a wrestler or an athlete) to enter World Wrestling Federation's thirty-man "Royal Rumble" match - which he entered to promote an improv comedy pay-per-view at the time. He appeared in a few backstage segments before his brief participation in the match. Upon entering the ring Carey stood unopposed for more than half a minute - he then eliminated himself by offering money to Kane and fleeing the ring.

Carey is known for his support of libraries, crediting them for beginning his successful comedy career. On May 2, 2000, in a celebrity edition of *Who Wants to Be a Millionaire*, he selected the Ohio Library Foundation to receive his $500,000 winnings. He later went on to win an additional $32,000 on the second celebrity *Millionaire*, making him one of the biggest winning contestants on *Millionaire* who did not win the top prize. Carey also has played on the World Poker Tour in the Hollywood Home games for the Cleveland Public Library charity. In June 2007 he offered to donate up to $100,000 (in $10,000 increments) to the Mooch Myernick Memorial Fund if anybody could beat him at the video game FIFA Soccer '07 for the Xbox 360. He dared five players from both the U.S. Men's and Women's National Teams to compete against him, and ended up donating $100,000 plus $60,000 for losing two games out of the six he played.

MACDONALD CAREY
Days of Our Lives

Edward Macdonald Carey (March 15, 1913 - March 21, 1994) was an actor best known for his role as the patriarch Dr. Tom Horton on NBC's soap opera *Days of Our Lives*. For almost three decades he was the show's central cast member.

Born in Sioux City, Iowa, Carey graduated from the University of Iowa with a bachelor's degree in 1935, after first attending the University of Wisconsin-Madison for a year where he was a member of Alpha Delta Phi. He first made his career starring in various B-movies of the 1940s, 1950s and 1960s, and was known in many Hollywood circles as "King of the Bs," sharing the throne with his "queen," Lucille Ball.

A successful radio actor and stage performer, his credits included the hit Broadway show *Lady in the Dark.* In 1942 Carey portrayed a Marine pilot who sank a Japanese cruiser in the WWII movie *Wake Island*, and in 1943 he joined the Corps for real and stayed in uniform for four years. After boot camp Carey was selected for Officer's Training School and commissioned a second lieutenant, and after training as a flight controller and ordnance officer he was deployed to Bougainville and the Philippines with Air Warning Squadron 3before mustering out of the Corps in 1947.

In 1943 he appeared in Alfred Hitchcock's *Shadow of a Doubt*, and returned to Paramount in 1947 with *Suddenly, It's Spring*. He continued with Paramount into the 1950s, and by that time had slipped into more noticeable character roles and had transitioned to westerns such as *The Great Missouri Raid, Outlaw Territory* and *Man or Gun*. Carey also played patriot Patrick Henry in *John Paul Jones*, and appeared in *Blue Denim*, *The Damned* (known as *These Are the Damned* in the US), *Tammy and the Doctor*, and *End of the World.*

In 1956 Carey took over the role of the kindly small-town physician Dr. Christian for one season, a character created in the late 1930s by the Danish-American actor Jean Hersholt, who had performed the part on radio and in films and had co-written a Dr. Christian novel. Carey also played the starring role of crusading Herb Maris in the 1950s syndicated series *Lock-Up.*

For the remainder of his career he played Tom Horton on *Days of our Lives.* During that time Carey suffered from a drinking problem, and eventually joined Alcoholics Anonymous in 1982. A longtime pipe smoker, he was seen with one in many films and early episodes of *Days of Our Lives*. He was ordered by his doctor to quit in September of 1991 after taking a leave of absence from *Days* in order to have a cancerous tumor removed from one of his lungs, and returned to the show in November of that year.

Carey is most recognized today, over a decade after his passing, as the voice who recites the epigram each day before the program begins: "Like sands through the hourglass, so are the days of our lives." From 1966 to 1994 he would also intone, "This is Macdonald Carey, and these are the days of our lives." (After his passing the producers, out of respect for Carey's family, decided not to use the second part of the opening tagline). At each intermission his voice also says "We will return for the second half of *Days of Our Lives* in just a moment." Since the Horton family is still regarded as the core of *Days*, his memory has been allowed to remain imprinted on the show by having the voiceovers remain intact. He also served as the voice-over for the very first PBS station identification in which he said "This is PBS...The Public Broadcasting Service."

Macdonald Carey wrote several books of poetry and a 1991 autobiography, *The Days of My Life,* and for his contribution to television he has a star on the Hollywood Walk of Fame at 6536 Hollywood Boulevard.

Carey was married to Elizabeth Hecksher from 1943 until their divorce in 1969. They had six children. Later he dated Lois Kraines, who was his significant other from 1973 on.

Macdonald Carey died in Beverly Hills from lung cancer in 1994 and is buried at Holy Cross Cemetery in Culver City, California alongside his daughter Lisa, who died in infancy. Carey also had five other children: Lynn, Steven, Edward Macdonald Jr., Paul, and Theresa, who is the mother of *Survivor: Panama Exile Island* winner Aras Baskauskas. Lynn Carey was a 70's Penthouse Pet and a well respected singer who provided music for Russ Meyer's legendary cult classic film *Beyond the Valley of the Dolls.*

BARRY CORBIN
Northern Exposure

Leonard Barrie "Barry" Corbin is an actor who was born in Lamesa, Texas on October 16, 1940 to a successful lawyer and an elementary school teacher. At the age of twenty-one Barry left Texas Tech University to join the Marine Corps on a friend's dare while suffering from a hangover. "We went in together," says his brother Blaine. "I worried about him. He wasn't the military type at all." Barry spent about two years at Camp Pendleton in California training South Vietnamese officers.

With no plans to abandon the Lone Star State, Barry joined the Marine Corps Reserve in March of 1962. He was initially attached to the 40th Rifle Company at Lubbock, and in boot camp was a member of the 3rd Recruit Training Battalion, Marine Corps Recruit Depot, San Diego. After completing his training in June of that same year he was ordered to Company N, 2nd Battalion, 2nd Infantry Training

Regiment at Marine Corps Base Camp Pendleton. He stayed there until he was released from active duty in September.

Barry remained in the Marine Corps Reserve, rejoining the 40th Rifle Company in Lubbock as an assistant Browning Automatic Rifle (BAR) man, and was discharged from the Reserves in August of 1963.

Barry still maintains that although he never left California, much less saw any action, his Marine Corps training has served him well in both his public and private pursuits. After his discharge he returned to Texas to pursue his dreams and started acting in regional theatres.

Corbin got bit by the performing bug at a young age when he would organize neighborhood plays. Later, he was enthralled by Westerns. He studied theatre at Texas Tech, where he performed everything from the great masters to contemporary playwrights.

In perhaps his best known role Barry portrayed Maurice Minnifield, town patriarch of Cicely, on the hit television show *Northern Exposure*. As a burly ex-astronaut, gung-ho president of the Cicely Chamber of Commerce, and owner of the town's newspaper and radio station, Maurice saw Cicely as a haven of limitless potential, soon to be the new "Alaskan Riviera." He also felt it was his duty to keep Dr. Fleischman (Rob Morrow) practicing in Cicely to secure the town's future urbanization.

He moved to New York in 1964 and during the next decade starred on Broadway, Off-Broadway, and in regional and dinner theaters in such roles as Henry in *Henry V*, Jud in *Oklahoma*, Oscar in *The Odd Couple,* Falstaff in *The Merry Wives of Windso*r, Henry in *Beckett*, and Macbeth in *Macbeth*.

Corbin relocated to Los Angeles in 1977. He was writing plays for National Public Radio when he was cast as Uncle

Bob in the feature film *Urban Cowboy*. He continued to create memorable   performances in films such as *The Man Who Loved Women, Nothing in Common, War Games, Best Little Whorehouse in Texas, Honky Tonk Man*, and John Hughes' *Career Opportunities*. Corbin's television credits include numerous miniseries and television movies such as *Lonesome Dove, The Thorn Birds, Fatal Vision, A Death in California, Last Flight Out, Young Harry Houdini*, and *Better Harvest*.

In 1996 Corbin returned to the stage with the one-man play *Charlie Goodnight's Last Ride*, which he co-wrote with Cowboy/Poet-singer Andy Wilkinson. He has also done voice work for commercials, radio, several books-on-tape and video games. Outside of acting, Barry is a cowboy and enjoys riding for celebrity events and on his own.

Corbin has served as the national spokesperson for National Alopecia Areata Foundation. He was diagnosed with it himself, and is especially passionate about helping children with alopecia. Alopecia areata (al-oh-PEE-shah air-ee-AH-tah) is a highly unpredictable, autoimmune skin disease resulting in the loss of hair on the scalp and elsewhere on the body.

During the run of *Northern Exposure* Corbin was reunited with a daughter he didn't know about. The actor discovered in late June 1991 he had a twenty-six-year-old daughter when Shannon Ross, who was adopted as an infant, tracked down her biological parents. Ross' mother had given her baby up for adoption at San Antonio's Methodist Mission Home in February of 1965 without telling Corbin she was pregnant with his child. She is seen dancing with him at the end of the episode *Midnight Sun*.

Corbin continues to act in film and television. In 2002 he appeared in the independent film *Waitin' to Live*, directed by

Joey Travolta (John's brother), and in 2003 he appeared in two television films, *Monte Walsh* and *Hope Ranch.*

Corbin won a Buffalo Bill Cody Award for quality family entertainment and the Western Heritage Award from the National Cowboy Hall of Fame for his performance in *Connagher.* He was also nominated for two Emmy Awards for Outstanding Supporting Actor in a Drama Series in 1993 and 1994 for *Northern Exposure,* a Media Owl Award, and an American Television Award for his work in *Northern Exposure.*

In 2003 Corbin returned to TV drama as Whitey Durham in the WB's *One Tree Hill.* Whitey is the long-time high school basketball coach in the town of Tree Hill.

He also reunited with actor John Cullum (*Holling*) in an award-winning independent short film, *Blackwater Elegy.* Corbin and Cullum play old friends who come to terms with their life following the death of a friend.

In the summer of 2004 Corbin filmed the role of Sheriff Buster Watkins in the feature *River's End* (previously titled *Molding Clay*), directed by William Katt. Filmed on location in Central, South and West Texas, Corbin plays a fictional Menard County sheriff who uses country savvy and cowboy logic to straighten out his angry teen-aged grandson, Clay, a high school senior who can't seem to stay out of trouble.

In 2006 Corbin appeared in *Beautiful Dreamer.* The film tells the story of a World War II fighter pilot who is shot down over Europe and declared dead. Two years later his wife finds him in a small town, but he doesn't remember her.

Corbin's distinctive voice has been lent to advertising over the last several years, most recently for Econo Lodge motels. He also narrates many projects, including the *A Fair to Remember: State Fair of Texas* documentary.

Corbin appeared in four episodes of *The Closer*, starting with the August 20, 2007 episode, as Brenda Johnson's (Kyra Sedgewick) father. He also had a couple of films released in 2007: *No Country for Old Men, Lake City, In the Valley of Elah*, and the short films *Trail End* and *A Death in the Woods.*

When not on a film or television set Corbin raises horses and cattle on his ranch in Fort Worth, Texas where he lives with daughter Shannon and his grandchildren. He also has three sons: Bernard, Jim, and Chris. He and his second wife, Susan, divorced in 1992.

Corbin appears at many roping and charity events, and recently helped the city of Lubbock, Texas celebrate its one hundred year anniversary.

DON CORNELIUS
Souuul Train!

Donald Cortez "Don" Cornelius is a television show host and producer who was born on September 27, 1936 in Chicago, Illinois He is best known as the creator of the nationally syndicated dance/music franchise, *Soul Train*, which he also hosted from 1971-1993.

Cornelius grew up on Chicago's predominantly black South Side and attended DuSable High School, where he studied art and drew cartoons for the school newspaper. After graduating from DuSable in 1954 he joined the Marines and served on a Korean airbase for eighteen months.

Upon returning to Chicago in 1956 Cornelius married his childhood sweetheart, Delores Harrison. The young couple soon had two children - Anthony in 1958, and Raymond in 1959 - and money was too short for Cornelius to afford college. "I needed money to support my family," he told *Billboard*, "so I sold tires, cars, and insurance."

Cornelius did well as a salesman - by 1966 he was making two-hundred-fifty dollars a week - but the arts tempted him. He wanted to become a radio announcer. "People always talked about my voice," he recalled in *Billboard*, "so I took a broadcasting course as a lark."

The four-hundred dollars Cornelius paid for the course was money well spent, because in 1966 he landed a part-time job as an announcer with WVON in Chicago. His take-home pay wasn't quite what he was used to - only one-hundred dollars a week - but he was fulfilling an ambition. "I started as a newsman, but I was also the swing (overnight) man," he explained. "I filled in as an all-around substitute at WVON. I felt I had to justify my job there... it was a very black station. I was sitting in for DJs and news people, doing public affairs outside the station, doing the talk show, and doing commercials." The job was a whirlwind, and the constant schedule changes began to wear on Cornelius, so when Roy Wood - Cornelius' superior and mentor at WVON - moved to a small UHF television station called WCIU-TV young Don began moonlighting for his former boss.

When Cornelius joined WCIU he already had *Soul Train* in mind. He took the name from a traveling music show he had hosted for WVON. The format - featuring dancing teenagers and popular records - came from Dick Clark's *American Bandstand*. What made *Soul Train* different was its black music format. WCIU's management was already attempting "ethnic" programming, so when Cornelius pitched his idea they agreed to give it a try… and the rest, as they say, is history.

BRIAN DENNEHY
Ned "Frozen Chosen" Coleman

Brian Mannion is a two-time Tony Award-winning actor who has appeared in movies, on television, and in live theater.

Dennehy was born in Bridgeport, Connecticut on July 9, 1938, the son of Hannah and Edward Dennehy, who was a wire service doctor for the Associated Press. He has two brothers, Michael and Edward. The family relocated to Long Island, where Dennehy attended Chaminade High School in the town of Mineola.

Rather than immediately chase his dreams of stage and screen Dennehy enlisted in the Marine Corps in 1959, actively serving until 1963. He went on to attend Columbia on a football scholarship to major in history, where he also became a member of the Sigma Chi fraternity before moving on to Yale to study dramatic arts. He played rugby for Old Blue RFC.

Dennehy is primarily known as a dramatic actor. His breakthrough role was as overzealous Sheriff (and decorated Marine combat veteran) Will Teasle in *First Blood* (1982) opposite Sylvester Stallone as Rambo. His other roles include a corrupt sheriff in the western *Silverado* and an alien in *Cocoon*, both released in 1985. He later played memorable supporting parts in such films as *Legal Eagles*, *F/X - Murder By Illusion*, *Presumed Innocent* and *F/X2 - The Deadly Art Of Illusion*.

During the 1980s Dennehy gradually became a valuable character actor in films and subsequently gained leading man status in the thriller *Best Seller* co-starring James Woods. He gained his art house spurs when he starred in the Peter Greenaway film *The Belly of an Architect*, for which he won the Best Actor Award at the 1987 Chicago International Film Festival. Commenting upon this unusual venture, Dennehy said, "I've been in a lot of movies, but this is the first *film* I've made."

Perhaps one of his most well known roles was in the 1995 Chris Farley-David Spade comedy *Tommy Boy* as Big Tom Callahan. Two of his earliest roles were in *10* with Bo Derek and Dudley Moore and *Foul Play* with Chevy Chase. Later on he would again star with Bo Derek in *Tommy Boy*. He also had a role in the recent movie *Ratatouille* as Django, Rem's Father.

Dennehy began his professional acting career is small guest roles in such 1970s and 1980s television series as *Kojak, Lou Grant, Dallas* and *Dynasty*, and appeared in an episode of *Miami Vice* during the 1987–88 season. Dennehy portrayed Marine Sergeant Ned T. "Frozen Chosen" Coleman in the 1980 television movie *A Rumor of War* opposite Brad Davis, and continued to appear in such high profile television movies as *Skokie, Day One, A Killing in a Small Town* with Barbara Hershey, *In Broad Daylight* and Scott Turow's *The Burden of Proof*. He also played a convincing Jackie Presser in HBO's *Teamster Boss: The Jackie Presser Story.*

He also had a lead role as fire chief/celebrity dad Leslie "Buddy" Krebs in the short-lived 1982 series *Star Of The Family.* Despite his notoriety, that show was cancelled after only two seasons.

Dennehy was nominated for Emmy Awards six times for his television movies, including one for his performance as John Wayne Gacy - for which he was nominated for Outstanding Lead Actor in a Miniseries or TV Movie. He was also nominated that same year in a different category, Outstanding Supporting Actor in a Miniseries or TV Movie, for *The Burden of Proof.* He has since been nominated for Emmy Awards for his work in *A Killing in a Small Town*, *Murder in the Heartland* and most recently for the Showtime cable TV movie *Our Fathers*, which was about the Roman Catholic Church sex abuse scandal.

In 2000 Dennehy was nominated for an Emmy for Outstanding Lead Actor in a Miniseries or TV Movie for a television presentation of his performance as Willy Loman in Arthur Miller's *Death of a Salesman,* which he had performed on Broadway. Although he did not win the Emmy (he has yet to win one), he did receive a Golden Globe award for the presentation.

Dennehy has won two Tony Awards, both times for Best Lead Actor in a Play. The first win was for *Death of a Salesman* (for which he also won a Laurence Olivier Award for the production's London run) in 1999, and the second was for Eugene O'Neill's *Long Day's Journey into Night* in 2003. Both productions were directed by Robert Falls and were originally produced at the Goodman Theatre company in Chicago.

On stage Dennehy has made frequent performances in the Chicago theatre world, and made his Broadway debut in 1995 in Brian Friel's *Translations*. In 1999 he was the first male performer to be voted the Sarah Siddons Award for his work in Chicago theatre. He made a return to Broadway in 2007 as Matthew Harrison Brady in *Inherit the Wind* opposite Christopher Plummer.

In 1989 Dennehy told the *New York Times* that he had received shrapnel wounds in the Vietnam War, and in 1993 he told *Playboy* that he had served five years in Vietnam. It was revealed, however, that he never served in Vietnam at all. In actuality he served during peacetime in the Marine Corps from 1959-1963, with Okinawa being his only overseas service. In 1999 Dennehy apologized for the fabricated stories.

At one point Dennehy resided at West Gilgo Beach, Long Island, and is currently a resident of Woodstock, Connecticut. He is the father of actresses Elizabeth and Kathleen Dennehy.

BRADFORD DILLMAN
Are You Anybody?

Bradford Dillman is a retired film and television actor and author who was born on April 14, 1930 in San Francisco, California to Dean and Josephine Dillman. He graduated from Yale University with a B.A. in English Literature, and following that served as a Marine in Korea from 1951 to 1953 before focusing on acting as a profession.

Studying with the Actor's Studio, he spent several seasons apprenticing with the Sharon, Connecticut Playhouse before making his professional acting debut in *The Scarecrow* in 1953. In 1956 Dillman took his initial Broadway bow in the Eugene O'Neill play *Long Day's Journey Into Night*, originating the author's alter ego character Edmund Tyrone and winning a Theatre World Award in the process. This distinct success put him squarely on the map, and 20th Century Fox took notice by placing the darkly handsome up-and-comer under contract. Cast in the melodramatic soap *A Certain Smile* in 1958, he earned a Golden Globe award.

After his debut in *A Certain Smile*, which co-starred Rossano Brazzi and Joan Fontaine, he appeared in many movies throughout the years including *Compulsion*, for which he won an award at Cannes, *A Circle of Deception*, the title role in *Francis of Assisi, Sanctuary, The Mephisto Waltz, Escape from the Planet of the Apes, The Iceman Cometh, The Way We Were, The Enforcer, The Swarm, Piranha* and *Lords of the Deep*. He also appeared on television throughout his career, starting on NBC's *Kraft Television Theatre* in 1954 and making a final acting appearance on *Murder She Wrote* in 1995. He also had a secondary but notable role in *The Bridge at Remagen* as the battalion commander of a mechanized infantry unit which seizes the Remagen bridges before it is destroyed by the Germans.

Bradford Dillman met actress and model Suzy Parker during the filming of *Circle of Deception*. They were married on April 20, 1963 and had three children, Dinah, Charles, and Christopher. The marriage lasted until her death on May 3, 2003. He was previously married to Frieda Harding from 1956 to 1962, and had two children (Jeffrey and Pamela) with her.

Dillman also wrote the football fan book *Inside the New York Giants* and the autobiography *Are You Anybody?: An Actor's Life*.

DALE DYE
Daddy Dye

Dale Adam Dye is an actor, presenter, businessman, and retired Marine who was born on October 8, 1944 in Cape Girardeau, Missouri, the son of Dale Adam Dye and Della Grace (née Koehler). Dye graduated from the Missouri Military Academy as an Officer Cadet and, lacking money for college, joined the Marine Corps in January of 1964. He was sent to Vietnam as a Marine Combat Correspondent from 1964-1965 and again from 1967-1970, surviving thirty-one major combat operations. During the war he received a Bronze Star and three Purple Hearts for wounds suffered in combat.

Dye spent thirteen years as an enlisted Marine, rising to the rank of Master Sergeant. He was chosen to attend Officer Candidates School and was appointed a Warrant Officer in 1976. He later converted his commission and was made a Captain. Dye was well-known in the tight-knit community of the Marine Combat Correspondents in Vietnam. It was fellow Marine correspondent Gustav Hasford who dubbed him "Daddy D.A." (since he was among the oldest of the correspondents) and included him as a character in his first semi-autobiographical Vietnam novel *The Short-Timers*, and even more extensively in his second, *The Phantom Blooper.* The movie based on Hasford's first novel, *Full Metal Jacket,* included the "Daddy D.A." character, (played by Keith Hodiak) though neither the character nor Dye's name is explicitly mentioned in the dialogue.

In his book *Dispatches*, journalist Michael Herr provided a vivid picture of Dye during the chaos of the Tet Offensive and the Battle of Hué. "And there was a Marine correspondent, Sergeant Dale Dye, who sat with a tall yellow flower sticking out of his helmet cover, a really outstanding target. He was rolling his eyes around and saying, 'Oh yes, oh yes, Charlie's got his shit together here, this will be bad,' and smiling happily. It was the same smile I saw a week later when a sniper's bullet tore up a wall two inches above his head, odd cause for amusement in anyone but a grunt."

After serving as a Captain in the Beirut Peacekeeping Force in 1982-83, Dye served in a variety of positions and got his B.A. in English from the University of Maryland University College. Then from 1983-84 he worked for the magazine *Soldier of Fortune* in Central America as he trained troops in guerrilla warfare in the countries of El Salvador and Nicaragua.

After his retirement from the Marines in 1984 Dye founded *Warriors, Inc.*, a company that specializes in training actors to realistically portray soldiers in movies of the war genre. In the 1986 movie *Platoon* he played Captain Harris, and also served as military technical advisor for the movie.

Dye has also appeared in several other films on which his company has advised. He played a role in the movie *Casualties of War,* and also played Colonel Robert F. Sink in the HBO miniseries *Band of Brothers.* He had a small role in *Saving Private Ryan* as an aide to General George Marshall, as well as a role playing the Admirals' aide Captain Garza in *Under Siege.* He has another small role in *Spy Game* as Commander Wiley during the rescue sequence, in *Mission Impossible* as Frank Barnes of the CIA, in *JFK* as General Y, and in *Starship Troopers* as a high-ranking officer in the aftermath of the Brain Bug capture. ("What's it thinking, Colonel?").

In addition to his on-screen work Dye hosts a Sunday evening radio show on KFI AM 640 Los Angeles, has been involved in the *Medal of Honor* series of video games as a consultant, and has hosted The History Channel's documentary series *The Conquerors.*

Dye's most recent project is an HBO companion piece to *Band of Brothers,* a ten-part mini-series known as *The Pacific War,* which was shot in Australia with Captain Dye working as a unit director shooting the Marines' major battles in the Pacific including Guadalcanal, Cape Gloucester, Peleliu, Iwo Jima and Okinawa.

DAVID EIGENBERG
Sex and the City

David Eigenberg is an actor who was born on Long Island, New York on May 17, 1964. He is best known for his role of Steve Brady on the HBO comedy *Sex and the City*.

Eigenberg's family moved to Illinois when he was four, and moved all around the suburbs of Chicago. Every summer from the age of twelve on was a working summer, saving for a college education that only lasted five weeks - before he was kicked out.

Acting was somewhat of a natural progression. Working for the first time at the age of twelve in a community theater project he landed a part, not in some cutesy kid story or as a clown, but in a production of Kurt Vonnegut's *Happy Birthday Wanda June*. The piece was pretty risqué for a small town, and walkouts abounded due to language and subject matter. Receiving his first and only good review, the local critic was very taken by David's presence. A few more

63

little local plays and musicals followed, but time and adolescence took him to different places.

Moving back to Naperthrill, he moved into an apartment above an abandoned car dealership and took several different jobs - driving auto parts, doing road construction, and finally enlisting in the Marine Corps Reserve and letting that wash over him. The intense mandate laid before all recruits, that they must learn to toe the line and suck it up and learn respect first for others and then themselves, had a deep impact on him and it resonated in all aspects of his life. He served for three years (1982-1986) and was Honorably Discharged at the rank of Lance Corporal.

Eigenberg applied for and was accepted to the American Academy of Dramatic Arts, and that decision took him to New York City. They were good days in the city, filled with a pace and energy he could now relate to - but it was years and years before anything of note came his way. Most of that time was spent in East Harlem, paying the bills with carpentry jobs and on paint crews and then stumbling into a friendship with Al Noccella - his partner in construction, and a beneficiary who kept him employed and then ushered him out when it looked like the 'break' had come. The character of Steve on *Sex and the City* came after many auditions for many parts on the show, and lasted for almost five years.

R. LEE ERMEY
Full Metal Jacket

"I don't have any respect at all for the scum-bags who went to Canada to avoid the draft or to avoid doing their fair share." – *R. Lee Ermey*

Ronald Lee Ermey (born March 24, 1944) is a former Marine Corps drill instructor and later Golden Globe-nominated actor who often plays the role of an authority figure such as Gunnery Sergeant Hartmann in *Full Metal Jacket*, Mayor Tilman in *Mississippi Burning* and Sheriff Hoyt in *The Texas Chainsaw Massacre*. He currently hosts *Mail Call*, a military history program on *The History Channel*, where Ermey answers military-related viewer questions. Ermey is also an official spokesman for Glock Firearms, Tupperware, Hoover, and the Young Marines, and has also appeared in commercials for Coors Light and Dick's Sporting Goods.

Born in Emporia, Kansas, Ermey enlisted in the Marine Corps in 1961 after being arrested several times as a teenager. A court judge gave him the choice of the military or jail. He later joked that the Marine Corps "put a screeching halt to my unconventional manner." He spent two years as a drill instructor at the Marine Corps Recruit Depots in San Diego, California and Parris Island, South Carolina from 1965 to 1967. In 1968 Ermey arrived in Vietnam where he served for fourteen months with the Marine Wing Support Group 17. He then served two tours of duty on Okinawa, Japan, during which he rose to the rank of Staff Sergeant and was medically discharged in 1972 for several injuries incurred during his tours. He did not receive a Purple Heart due to his injuries being noncombat in origin.

Ermey was cast in his first movie while attending the University of Manila in the Philippines using his G.I. Bill benefits. He first played a Marine drill instructor (SSgt Loyce) in the 1978 Vietnam-era film *The Boys in Company C*, which brought Ermey to the attention of director Stanley Kubrick. He then played an Air Cavalry Officer in *Apocalypse Now*, doubling as a technical advisor to director Francis Ford Coppola on that film. For the next few years Ermey played a series of minor film roles until 1987, when he was cast as tough drill instructor Gunnery Sergeant Hartmann in Stanley Kubrick's *Full Metal Jacket*. Ermey also served as the technical advisor on the film. He was originally only intended to be the technical advisor, but Kubrick changed his mind after Ermey put together an instructional tape to convince Kubrick he was the right person for the role, during which he went on an extended hair-raising drill instructor tirade while being pelted by oranges and tennis balls - all without repeating himself, stopping, or even flinching. Kubrick allowed him to write his own dialogue and improvise on set, a noted rarity in a Kubrick film. Kubrick later indicated that Ermey was an excellent performer, needing just two or three takes per scene, also a noted rarity for a Kubrick film. Ermey was so convincing on set as the DI that on one occasion he barked an order at Kubrick, who instinctively stood at attention and followed orders before realizing what had happened. Ermey's performance won critical raves and he was nominated for a Golden Globe Award as Best Supporting Actor. He would subsequently play a tough drill instructor in the pilot episode of *Space: Above and Beyond* and the ghost of a drill instructor in the film *The Frighteners*, both similar to his character in *Full Metal Jacket.*

He has since appeared in approximately sixty films, including *Mississippi Burning, Dead Man Walking, Seven, Leaving Las Vegas, Prefontaine, Saving Silverman, On Deadly Ground, Life, Man of the House*, and *Toy Soldiers*, as well as the remakes of *Willard* and *The Texas Chainsaw Massacre.* Ermey has also lent his distinctive voice to *The Grim Adventures of Billy & Mandy, Toy Story* and *Toy Story 2*, as well as *Roughnecks* and *X-Men 3*.

Ermey has long been a major advocate for troops overseas and traveled to Kuwait in June of 2003 during the first phase of Operation Iraqi Freedom to film mail distribution by the Defense Department to service personnel for an episode of *Mail Call*. He has also conducted morale tours visiting U.S. troops in locations such as Bagram Airbase, Afghanistan in which he filmed segments for *Mail Call*. While at Bagram he held a USO type show in which he portrayed GySgt Hartman and conducted a comedy routine. He also did the same thing at Doha, Qatar in 2003.

According to a 2005 episode of *Mail Call* filmed at Whiteman Air Force Base, Ermey is the 341st person to fly in the B-2 Stealth Bomber, and on May 17, 2002 he received an honorary promotion to Gunnery Sergeant from the Commandant of the Marine Corps (later Supreme Allied

Commander Europe (SACEUR)) General James L. Jones, becoming the first retired military member in the history of the Marine Corps to be promoted.

On *Mail Call* Ermey discusses weaponry, tactical matters, and military history. *Mail Call's* subject matter is dictated by viewer emails - one episode may focus on an M1A1 Battle Tank, while another may involve World War II secrets, while a third might focus on elements of medieval warfare. The set consists of a military tent, other military gear and weapons, and Ermey's personal jeep armed with his own .30 cal. M1919 Browning machine gun. Commercial breaks are signaled with typical DI type language such as, "Goin' to the can? I don't *think* so, sweet pea. Keep your butt parked on that couch!"

MIKE FARRELL
*M*A*S*H*

"I grew up Catholic, and as they say, once a Catholic, always a Catholic... just like I was a Marine, and once you've been a Marine, you're always a Marine."
- *Mike Farrell*

Mike Farrell (born February 6, 1939) is an actor best known for his role as Captain B.J. Hunnicutt on the popular television series *M*A*S*H* (1975–83). More recently, Farrell has starred on the television series *Providence* (1999–2002) and appeared as Milton Lang, Victor's father, on *Desperate Housewives (*2007–2008). He is also a prominent activist for politically left-wing causes.

Farrell was born in St. Paul, Minnesota as one of four children. When he was two years old his family moved to Hollywood, California, where his father worked as a movie studio carpenter. Farrell attended West Hollywood Grammar School with Natalie Wood, graduated from Hollywood High

School, served in the Marine Corps, and worked at various jobs before his acting career.

During the 1960s Farrell guest-starred on a few series. Notable roles included 'Federal Agent Modell' in an episode of *The Monkees* in 1967, and astronaut Arland in an episode of *I Dream of Jeannie*. In 1968 he originated the continuing role of Scott Banning in the NBC soap *Days of Our Lives*, and in 1970 he starred as one of the young doctors on the CBS prime-time series *The Interns* in a cast led by Broderick Crawford. In 1971 he played the assistant to Anthony Quinn on ABC's *The Man and The City* ,and in 1973, while under contract to Universal Studio, Farrell starred with Robert Foxworth on *The Questor Tapes*.

Farrell's big break came in 1975 when Wayne Rogers unexpectedly departed *M*A*S*H* at the end of the third season, and Farrell was quickly recruited for the newly created role of B.J. Hunnicutt. He stayed with the series for its remaining eight years on the air. During that time, Farrell wrote five episodes and directed four.

Besides being a writer and a director, Farrell has been an executive producer and a producer in both television and film including *Dominick and Eugene* and *Patch Adams*.

In 1999 he was given the part of veterinarian Jim Hanson (the father of the lead character, Dr. Sydney Hansen, portrayed by actress Melina Kanakaredes) on the NBC-TV melodrama series *Providence*. In his portrayal of Sydney's father, Farrell played opposite actress Concetta Tomei, who portrayed his wife (Lynda Hansen). Tomei's character died during the first episode of the series, but continued to appear as a ghost/memory in later episodes. The show would prove to be a big hit with the critics and in the Nielsen Ratings. Farrell appeared in 64 of the 92 episodes before its surprising cancellation in December of 2002.

Even before he was well-known, Farrell was an activist for many political and social causes. He has worked with Human Rights Watch, was on the Board of Advisors of the original Cult Awareness Network, and has been president of Death Penalty Focus for more than ten years.

In 1985 Farrell was in Central America, helping refugees from the civil war in El Salvador. A guerrilla commander, Nidia Diaz, had been taken prisoner. She needed surgery, but no Salvadoran doctor dared to help her. Amnesty International recruited a foreign doctor, and Farrell was present as an observer but was, in his words, "shanghaied into assisting with the surgery" when the doctor said his help was needed. The in-prison surgery was successful, and Diaz went on to be one of the signers of the Chapultepec Peace Accords (the peace treaty ending the war), and served in the Constituent Assembly of El Salvador and in the Central American Parliament.

Farrell has also been active in the Screen Actors Guild, and in 2002 he was elected First Vice-President of the Guild in Los Angeles. He served in the post for three years.

In 2006 Farrell appeared with Jello Biafra and Keith Gordon in the documentary *Whose War?*, examining the U.S. role in the Iraq War.

He married actress Judy Farrell in 1963, but they divorced in 1983. They have two children, Michael and Erin - and on *M*A*S*H* B.J's fictional daughter was also named Erin. On December 31, 1984 Farrell married actress Shelley Fabares.

GLENN FORD
Global Marine

"Let's never forget that to remain free we must always be strong... national defense must be the top priority for our country. If you are strong, you are safe. Now is the time for every American to be proud. This is the land of the free and the home of the brave. If we are not brave, we will not be free." - *Glenn Ford*

Gwyllyn Samuel Newton "Glenn" Ford (May 1, 1916 - August 30, 2006) was an acclaimed Canadian-born actor from Hollywood's Golden Era with a career that spanned seven decades. Ford was a versatile actor best known for playing either cowboys or ordinary men in unusual circumstances.

He was born at Jeffrey Hale Hospital in Quebec City, Quebec to Anglo-Quebecer parents Hannah and Newton Ford, who was a railroad executive, and was a great-nephew of Canada's first Prime Minister Sir John A. Macdonald. Ford moved to Santa Monica, California with his family at the age of eight, and became a naturalized citizen of the United States in 1939.

After Ford graduated from Santa Monica High School he began working in small theatre groups. He later commented that his railroad executive father had no objection to his growing interest in acting, but told him, "It's all right for you to try to act, if you learn something else first. Be able to take a car apart and put it together. Be able to build a house, every bit of it. Then you'll always have something." Ford heeded the advice and during the 1950s, when he was one of Hollywood's most popular actors, regularly worked on the plumbing, wiring, and air conditioning at home. At times, he even worked as a roofer and installer of plate-glass windows.

He acted in West Coast stage companies before joining Columbia Pictures in 1939. His stage name came from his father's hometown of Glenford, Canada. His first major movie part was in the 1939 film *Heaven with a Barbed Wire Fence.*

In December of 1942 Ford's film career was interrupted when he volunteered for duty with the Marine Corps Reserve as a photographic specialist at the rank of Sergeant. He was assigned in March 1943 to active duty at the Marine Corps Base in San Diego, and was later sent to Marine Corps Schools Detachment (Photographic Section) in Quantico, Virginia that June with orders as a motion-picture production technician. Sergeant Ford returned to the San Diego base in February 1944, and was assigned next to the radio section of the Public Relations Office, Headquarters Company, Base Headquarters Battalion. There he staged and broadcast the radio program *Halls of Montezuma.* Glenn Ford was honorably discharged from the Marines on 7 December 1944.

Fourteen years later, in 1958, he joined the Naval Reserve and was commissioned a lieutenant commander with a 1655 designator (public affairs officer). During his annual training tours he promoted the Navy through radio and television broadcasts, personal appearances, and documentary films. He was promoted to Commander in 1963, and Captain in 1968.

Ford went to Vietnam in 1967 for a month's tour of duty as a location scout for combat scenes in a training film entitled *Global Marine*. He traveled with a combat camera crew from the demilitarized zone south to the Mekong Delta. For his service in Vietnam, the Navy awarded him a Navy Commendation Medal. He retired from the Naval Reserve in the 1970s at the rank of Captain.

Following military service Ford's breakthrough role was in 1946, starring alongside Rita Hayworth in *Gilda*. He went on to be a leading man opposite her in a total of five films. While the movie is mostly remembered as the vehicle for Hayworth's "provocative rendition of a song called *Put the Blame on Mame*," The New York Times movie reviewer Bosley Crowther praised Ford's "stamina and poise in a thankless role" despite the movie's poor direction.

Ford's career flourished in the 1950s and into the 1960s, and continued into the early 1990s with an increasing number of television roles. His major roles in thrillers, dramas and action films include *A Stolen Life* with Bette Davis, *The Secret of Convict Lake* with Gene Tierney, *The Big Heat, Framed, Blackboard Jungle, Interrupted Melody, Experiment in Terror, Four Horsemen of the Apocalypse, Ransom!, Superman* and westerns such as *The Fastest Gun Alive, 3:10 to Yuma* and *Cimarron*. Ford's versatility also allowed him to star in a number of popular comedies, including *The Teahouse of the August Moon, Don't Go Near the Water, The Gazebo, Cry For Happy,* and *The Courtship of Eddie's Father.*

In 1971 Ford signed with CBS to star in his first television series, a half hour comedy/drama titled *The Glenn Ford Show,* however CBS head Fred Silverman noticed many of the featured films being shown at a Glenn Ford film festival were westerns. He suggested doing a western series instead, which resulted in the "modern day western" series, *Cade's County.* Ford played southwestern Sheriff Cade for one season (1971-1972) in a mix of western drama and police mystery. In *The Family Holvak* (1975-1976), Ford portrayed a depression era preacher in a family drama, reprising the same character he had played in the TV film *The Greatest Gift.*

In 1978 Ford had a supporting role in *Superman* as Clark Kent's adopted father Jonathan Kent, a role that introduced him to a new generation of film audiences. Ford's final scene in the film begins with a direct reference to *Blackboard Jungle* - the earlier film's theme song *Rock Around the Clock* is heard on a car radio.

In 1991 Ford agreed to star in a cable network series called *African Skies,* however prior to the start of the series he developed blood clots in his legs which required a lengthy stay in Cedars-Sinai Medical Center. He eventually recovered, but at one time his situation was so severe that he was listed in critical condition. Ford was forced to drop out of the series, and was replaced by Robert Mitchum.

In the 2006 movie *Superman Returns* there is a scene where Ma Kent (played by Eva Marie Saint) stands next to the living room mantel after Superman returns from his quest to find remnants of Krypton. On that mantel is a picture of Pa Kent (as played by Glenn Ford). This "cameo" of sorts was Ford's last screen appearance (the photograph is more easily visible in a deleted scene included with the DVD release of the film).

Ford's first wife was actress and dancer Eleanor Powell (1943-1959), with whom he had his only child Peter, who was born in 1945. The couple appeared together on screen once in a short subject produced in the 1950s entitled *The Faith of Our Children.* Ford subsequently married actress Kathryn Hays (1966-1969), Cynthia Hayward (1977-1984) and Jeanne Baus (1993-1994). All four marriages ended in divorce. Ford was not on good terms with his ex-wives. He also had a long-term relationship with actress Hope Lange, although they never married.

In 1978 Ford underwent hypnosis at his home in Beverly Hills and recalled a past life as a Colorado cowboy named Charlie Bill. He gave a detailed description of the past, which was tape-recorded for academics at the University of California to study. A second experiment was conducted at the university itself when Ford, then sixty-one, responded well to the hypnosis. This time he did not recall the life of Charlie Bill, but that of a Scottish piano teacher named Charles Stuart. "I teach the piano to young flibbertigibbets," said Ford under the hypnosis, using a quaint old English word for rascals not in common use in California. He allegedly played a few notes on piano during the experiment, despite later saying he had never been taught to play the instrument. The researchers then managed to locate the grave of a Charles Stuart in Elgin, Scotland, who died in 1840. After being shown a photo of the burial place, Ford said "That shook me up real bad. I felt immediately that it was the place I was buried."

For the first half of his life Ford supported the Democratic Party - in the 1950s he supported Adlai Stevenson for President - and in later years became a supporter of the

Republican Party, campaigning for his friend Ronald Reagan in the 1980 and 1984 presidential elections.

After being nominated in 1957 and 1958, in 1962 Glenn Ford won a Golden Globe Award as Best Actor for his performance in Frank Capra's *Pocketful of Miracles*. He was listed in Quigley's Annual List of Top Ten Box Office Champions in 1956, 1958 and 1959, topping the list at number one in 1958. For his contribution to the motion picture industry Glenn Ford has a star on the Hollywood Walk of Fame at 6933 Hollywood Blvd, and in 1978 he was inducted into the Western Performers Hall of Fame at the National Cowboy & Western Heritage Museum in Oklahoma City. In 1987 he received the Donostia Award in the San Sebastian International Film Festival, and in 1992 was awarded the Légion d'honneur medal for his actions during the Second World War.

Ford was scheduled to make his first public appearance in fifteen years at a ninetieth birthday tribute gala in his honor hosted by the American Cinematheque at Grauman's Egyptian Theatre in Hollywood on May 1, 2006, but had to bow out at the last minute. Anticipating that his health might prevent his attendance, Ford had the previous week recorded a special filmed message for the audience which was screened after a series of in-person tributes from friends including Martin Landau, Shirley Jones, Jamie Farr, and Debbie Reynolds.

Glenn Ford died in his Beverly Hills home on August 30, 2006 at the age of ninety. In October of 2008 son Peter Ford held a live auction on the Internet to sell some of his father's possessions, including Ford's lacquered cowboy boots, a jacket and cap from *The White Tower*, wool trench coat from *Young Man with Ideas*, and his United States Naval Reserve uniform cap.

CHRISTOPHER GEORGE
Rat Patrol

Christopher John George (February 25, 1929 - November 28, 1983) was a television and film actor who was perhaps best known for his starring role in the 1966-1968 TV series *The Rat Patrol*. He was nominated for a Golden Globe in 1967 as Best TV Star for his performance in the series. George was also the long-time husband of actress Lynda Day George, and his niece is *Wheel of Fortune* hostess Vanna White.

George was born in Royal Oak, Michigan, the son of Greek immigrants. His father was a traveling salesman during his childhood, and his family eventually settled in Miami, where George attended Miami High School. He served in the Marines, and later attended the University of Miami. George began acting in New York City, where he performed on the stage and in television commercials.

George first rose to prominence in 1967 playing a supporting role in the Howard Hawks-directed western film

El Dorado, starring John Wayne. George and Wayne became friends while shooting the film, and would co-star in additional westerns including *Chisum* in 1970 and *The Train Robbers* in 1973.

From 1966 to 1968 George played the lead role of Sergeant Sam Troy in *The Rat Patrol*. The show followed the exploits of four allied soldiers who were part of a long range desert patrol group in the North African campaign during World War II. Following cancellation of the series he played the lead role in several genre films of the 1960s including *Tiger by the Tail*, Project X, and *The Devil's 8*. In 1969 George portrayed Ben Richards in the pilot movie for *The Immortal*, which ran as the ABC Movie of the Week. The film was picked up as a TV series and ran for fifteen episodes from 1970-1971. During this time he also portrayed Dan August in the television film *House On Greenapple Road*, which evolved into the 1970-71 series *Dan August* starring Burt Reynolds.

George first met actress Lynda Day when they starred together in the 1966 independent film *The Gentle Rain*. They would star together again four years later in *Chisum*, where they fell in love and soon married. Thereafter Lynda became Lynda Day George and co-starred with Christopher in multiple television films over the next ten years including *Mayday at 40,000 Feet* and *Cruise Into Terror*. They also worked together in episodes of *The F.B.I.*, *Mission: Impossible*, *McCloud*, *Wonder Woman*, *Love Boat* and *Vegas*.

George continued his television work throughout the 1970s with guest roles on many popular series. He also surprised fans by posing nude for *Playgirl* magazine in the June, 1974 issue. In 1976, George played a supporting role in the all-star World War II epic *Midway*. That same year he played the lead role of Ranger Michael Kelly in the Film Ventures International independent film *Grizzly*. A thinly-veiled *Jaws* clone, the animal horror thriller became one of the most popular films of Christopher's career, earning more than thirty-nine million at the box office.

George followed that success with a busy string of horror, action, splatter and slasher B-movies over the next seven years including *Dixie Dynamite* (co-starring fellow Marine Warren Oates), *Day of the Animals*, *Whiskey Mountain*, *City of the Living Dead*, *The Exterminator*, *Graduation Day*, *Enter the Ninja*, *Pieces*, and *Mortuary*. Many of these works have since achieved cult film status.

Christopher George died unexpectedly of a heart attack on November 28, 1983 at the age of fifty-four. A contributing factor in his death is believed to have been a 1967 mishap suffered on the set of *The Rat Patrol* when his jeep flipped over and pinned him beneath the vehicle, giving him a cardiac contusion. His death devastated wife Lynda, and afterwards she only worked sporadically in television guest roles until her retirement in the early 1990s. He is interred in the Westwood Village Memorial Park Cemetery in Los Angeles.

SCOTT GLENN
The Right Stuff

Theodore Scott Glenn (born January 26, 1941) is an actor known for a deadpan and serious delivery and demeanor. His roles include Wes Hightower in *Urban Cowboy*, astronaut Alan Shepard in *The Right Stuff*, Commander Bart Mancuso in *The Hunt for Red October*, and FBI Agent Jack Crawford in *The Silence of the Lambs*.

Glenn was born in Pittsburgh, Pennsylvania, the son of Elizabeth, a homemaker, and Theodore Glenn, a business executive. He grew up in Appalachia and has Irish and Native American ancestry. During his childhood he was regularly ill, and for a year was bed-ridden. Through intense training programs he got over his illnesses, including a limp. After graduating from a Pittsburgh high school Glenn entered the College of William and Mary where he majored in English. He then joined the Marines for three years, and worked roughly five months as a reporter for the *Kenosha Evening News*. He then tried to become an author, but found he could not write good dialogue. To learn the art, he began taking acting classes.

In 1965 Glenn made his Broadway debut in *The Impossible Years.* He joined George Morrison's acting class and helped to direct student plays to pay for his studies, and appearing onstage in *La MaMa Experimental Theatre Club* productions. In 1967 he married Carol Schwartz, his current wife, and converted to his wife's Jewish religion upon marrying her. In 1968 he joined *The Actors Studio* and began working in professional theatre and TV. In 1970 director James Bridges offered him his first movie role in *The Baby Maker*, released the same year.

Glenn left that year for LA and spent about eight years there acting in small film roles and doing brief TV stints, including a TV movie called *Gargoyles.* He appeared in Francis Ford Coppola's *Apocalypse Now* in a small role, and also worked with directors like Jonathan Demme and Robert Altman. Fed up with Hollywood, in 1978 Glenn left Los Angeles with his family for Ketchum, Idaho and for the two years he lived there worked as a barman, huntsman and mountain ranger, while occasionally acting in Seattle stage productions.

In 1980 Glenn got back into acting in films by appearing as ex-convict Wes Hightower in Bridges' *Urban Cowboy.* After that he appeared in the gothic horror film *The Keep,* action films like *Silverado* and *The Challenge,* drama films like *The Right Stuff,* alternately playing good guys and bad guys. He returned to Broadway in *Burn This* in 1987. That same year he tried his hand at gangster movies when he starred as the real-life sheriff turned gunman Verne Miller in the movie of the same name. *Verne Miller* was only given a theatrical release in Finland and went straight to video in the U.S. In the beginning of the 1990s his career was at its peak as he appeared in several well-known and/or blockbuster films such as *The Silence of the Lambs*, *Backdraft, The Hunt for Red October,* and *The Player*. He also played a vicious hitman in a critically acclaimed performance in *Night of the Running Man*. Later he gravitated toward more challenging movie roles such as in the Freudian farce *Reckless*.

Glenn's most recent theatrical roles were in *The Bourne Ultimatum* and the drama *Freedom Writers*, in which he played the father of Hilary Swank's character.

JAMES GREGORY
Inspector Luger

James Gregory (December 23, 1911 - September 16, 2002) was a character actor who was born in the Bronx, New York, and grew up in New Rochelle. He was noted for his deep, gravelly voice and for playing brash roles such as McCarthy-like Senator John Iselin in *The Manchurian Candidate*, Morgan Hasting (opposite John Wayne and Dean Martin) in *The Sons of Katie Elder*, the audacious gorilla General Ursus in *Beneath the Planet of the Apes*, and loudmouthed Inspector Luger in the *Barney Miller* TV series which ran from 1975 to 1982. He also played Dean Martin's spy boss MacDonald in the *Matt Helm* movie series, and is fondly remembered for his role as Dr. Tristan Adams, the villainous director of the Tantalus IV Penal Colony, on the original *Star Trek* episode *Dagger of the Mind*. Another role many recall was as the father of Scott Hayward in Elvis Presley's 1967 musical *Clambake*.

In high school Gregory was president of the Drama Club. He worked on Wall Street as a runner in 1929 and thought of being a stockbroker but, by 1935, had become a professional actor instead. In 1939 he made his Broadway debut in a production of *Key Largo,* and did about twenty-five more Broadway productions over the next sixteen years.

Gregory served three years in the Marine Corps and Navy during World War II, and his tour of duty took him to the Pacific where he spent eighty-three days on Okinawa. His early acting work included Army training films, and one such appearance is excerpted in *The Atomic Café.*

From 1959 to 1961 Gregory had his own NBC series, a 1920s crime drama entitled *The Lawless Years.* He played a New York City police detective named Barney Ruditsky, with Robert Karnes co-starring as Max Fields. He also starred in *PT 109* with Cliff Robertson in 1963.

James Gregory died in Sedona, Arizona on September 16, 2002 of natural causes at the age of ninety.

CLU GULAGER
The Last Picture Show

William Martin "Clu" Gulager is a television and film actor born November 16, 1928 in Holdenville, Oklahoma. He is the son of John Gulager, a cowboy entertainer, and his first cousin was Will Rogers. Early in life Clu served in the Marine Corps and was stationed at Camp Pendleton from 1946 to 1948. He learned his skills at an experimental theatre in Paris under the guidance of actor Jean-Louis Barrault, and then signed up with Universal as a contract player. He is particularly noted for appearing in 1980s horror movies such as *The Return of the Living Dead, The Hidden,* and *The Offspring* (aka *From a Whisper to a Scream*).

Gulager played Billy the Kid in the 1960 to 1962 NBC television series *The Tall Man* opposite Barry Sullivan as Pat Garrett, and Emmett Ryker from 1964 to 1968 on another NBC series called *The Virginian* which starred James Drury. In addition he appeared in more than sixty other roles in film and television. He also starred in the 1964 version of *The Killer,* and appeared notably in *The Last Picture Show* along with Cybil Shepherd and Ellen Burstyn.

Gulager is the father of film director John Gulager (who was the contest winner on third season of *Project Greenlight*), and appeared in his film *Feast* as a shotgun-toting bartender. He is the widower of actress Miriam Byrd-Nethery, who died in 2003.

GENE HACKMAN
Popeye Doyle

Eugene Allen "Gene" Hackman is an accomplished actor who was born on January 30, 1930 in San Bernardino, California to Lyda (née Gray) and Eugene Ezra Hackman. He came to fame during the 1970s after his role as 'Popeye' Doyle in *The French Connection,* and continued to appear in Hollywood films playing major roles, including Harry Caul in *The Conversation,* Norman Dale in *Hoosiers*, Agent Rupert Anderson in *Mississippi Burning*, Little Bill Daggett in *Unforgiven*, Lex Luthor in *Superman*, Captain Frank Ramsey in *Crimson Tide*, Joe Moore in *Heist* and Admiral Leslie McMahon Reigart in *Behind Enemy Lines*.

Hackman's family moved from one place to another until finally settling in Danville, Illinois. They lived in the house of his maternal grandmother, Beatrice, and Hackman's father operated the printing press for a local paper called the *Commercial-News*. Hackman's parents divorced in 1943, and his mother died in 1962 as the result of a fire she accidentally set while smoking. At sixteen Hackman left home to join the Marine Corps, where he served four-and-a-half years as a field radio operator. After finishing his service he moved to New York and worked in several minor jobs before moving on to study television production and journalism at the University of Illinois under the G.I. Bill.

At the age of twenty-six Hackman decided to become an actor and joined the Pasadena Playhouse in California. It was there that he forged a friendship with another aspiring actor, Dustin Hoffman. Already seen as outsiders by their classmates, Hackman and Hoffman were later voted "The Least Likely to Succeed." Determined to prove them wrong, he hopped on a bus bound for New York City. A 2004 article in *Vanity Fair* described how Hackman, Hoffman and Robert Duvall were all struggling actors and close friends while living in New York City in the 1960s. Hackman was working as a doorman when he ran into an instructor whom he had despised at the Pasadena Playhouse. Reinforcing "The Least Likely to Succeed" vote, the man said "See Hackman, I told you that you wouldn't amount to anything." That was all the motivation the former Marine needed.

Hackman soon began performing in several off-Broadway plays, and finally in 1964 he received an offer to co-star in the play *Any Wednesday* with actress Sandy Dennis. This opened the door to film work. His first role was in *Lilith*, with Warren Beatty in the leading role. Another supporting role, Buck Barrow in 1967's *Bonnie and Clyde*, earned him an Academy Award nomination as Best Supporting Actor.

Other roles followed. In 1969 he played a ski coach in *Downhill Racer*, and an astronaut in *Marooned*. Also In 1969 he played a member of a barnstorming skydiving team that entertained at county fairs. *The Gypsy Moths* is still considered by hard core skydivers & BASE jumpers to be the best movie on the Extreme Skydiving lifestyle.

In 1970 he was nominated for the same award, this time for *I Never Sang for My Father*, working alongside Melvyn Douglas and Estelle Parsons. The next year he won the Best Actor award for his memorable performance as New York City police officer Popeye Doyle in *The French Connection*, marking his graduation to leading man status. He followed this with leading roles in the disaster film *The Poseidon Adventure* and Francis Ford Coppola's *The Conversation*, which was nominated for several Oscars. That same year Hackman appeared uncredited in one of his most famous comedic roles as the blind hermit in *Young Frankenstein*, and later was cast as Polish General Sosabowski in the star-studded war film *A Bridge Too Far*.

By the end of the 1980s Gene Hackman was a well respected actor and alternated between leading and supporting roles, earning another Best Actor nomination for *Mississippi Burning*, and appearing in such films as *Reds, Under Fire, Hoosiers, Power, Bat*21* and *Uncommon Valor*. His performance as a rural Indiana high school basketball coach in the period drama *Hoosiers* is considered by some to be particularly memorable, and in *Uncommon Valor* he played a retired Marine Colonel who returns to Vietnam to find his MIA son.

In 1990 Hackman underwent heart surgery which kept him from work for a while, although he found time for *Narrow Margin* - a remake of 1952's *The Narrow Margin*. Then in 1992 he played the sadistic sheriff "Little Bill" Daggett in the western *Unforgiven*, which was directed by Clint Eastwood and earned him a second Oscar, this time for Best Supporting Actor. The film itself won Best Picture.

In 1995 Hackman played other noteworthy villains, such as fast-draw champion John Herrod in *The Quick and the Dead* opposite Leonardo DiCaprio and Russell Crowe and submarine Captain Frank Ramsey in the film *Crimson Tide* with Denzel Washington. In 1996 he took a comedic turn as ultra-conservative Senator Kevin Keeley in *The Birdcage* with Robin Williams and Nathan Lane, and later co-starred with Will Smith in the 1998 film *Enemy of the State* - where his character was reminiscent of the one he played in *The Conversation.*

In 2003, at the Golden Globe Awards, Hackman was honored with the Cecil B. DeMille Award for his "outstanding contribution to the entertainment field." He has also written three novels with undersea archaeologist Daniel Lenihan - *Wake of the Perdido Star, Justice for None, and Escape from Andersonville.*

On July 7, 2004 Hackman gave a rare interview to Larry King during which he announced he had no future film projects lined up, and that his acting career was over. His final film was *Welcome to Mooseport*, a comedy with Ray Romano in which Hackman portrayed a former President of the United States.

Hackman's first wife was Faye Maltese. They had three children, Christopher Allen, Elizabeth Jean, and Leslie Anne, but the couple divorced in 1986 after thirty years of marriage. Then in 1991 Hackman married Betsy Arakawa. They now live in Santa Fe, New Mexico where Betsy is co-owner of an upscale retail home-furnishing store called *Pandora's, Inc.*

STERLING HAYDEN
Blond Viking God

Sterling Hayden (March 26, 1916 - May 23, 1986) was an actor and author. For most of his career as a leading man he specialized in westerns and film noir such as *Johnny Guitar, The Asphalt Jungle* and *The Killing*. Later on he became noted as a character actor for such roles as General Jack D. Ripper in *Dr. Strangelove* and the Irish policeman, Captain McCluskey, in Francis Ford Coppola's *The Godfather*.

Born in Upper Montclair, New Jersey, Hayden's parents were George and Frances Walter, and named him Sterling Relyea Walter. After his father died he was adopted at the age of nine by James Hayden and renamed Sterling Walter Hayden, and as a child lived in New Hampshire, Massachusetts, Pennsylvania, Washington D.C., and Maine, where he attended Wassookeag School in Dexter.

Hayden was a genuine adventurer and man of action, not dissimilar from many of his movie parts. He ran away to sea at seventeen as a ship's boy, and later was a fisherman on the

Grand Banks of Newfoundland. After serving as sailor and fireman on larger vessels he was awarded his first command at nineteen, and sailed around the world several times.

Hayden became a print model and later signed a contract with Paramount Pictures, who dubbed the 6' 5" actor *The Most Beautiful Man in the Movies* and *The Beautiful Blond Viking God*. His first film starred Madeleine Carroll, with whom he fell in love and married. Then, after just two film roles, he left Hollywood to serve as an undercover agent with William J. "Wild Bill" Donovan's COI office and remained there after it became the OSS. Hayden also joined the Marine Corps under the name 'John Hamilton' (which was never his legal name). His World War II service included running guns through German lines to the Yugoslav partisans and parachuting into fascist Croatia. He was awarded the Silver Star, and received a commendation from Yugoslavia's Marshal Tito.

Hayden's admiration for the Communist partisans led to a brief membership in the Communist Party. As the Red Scare deepened in U.S. he cooperated with the House Un-American Activities Committee, confessing his brief Communist ties and 'naming names.' His wife at that time, Betty De Noon, insisted the 'names' her ex-husband provided were already in the hands of the Committee, which had a copy of the Communist Party's membership list. In any event Hayden subsequently repudiated his own cooperation with the Committee, stating in his autobiography "I don't think you have the foggiest notion of the contempt I have had for myself since the day I did that thing."

Sterling Hayden often professed distaste for film acting, claiming he did it mainly to pay for his ships and voyages. In 1959, after a bitter divorce, he was awarded custody of his children. He then defied a court order and sailed to Tahiti with all four - Christian, Dana, Gretchen and Matthew.

Hayden married Catherine Devine McConnell in 1960. They had two sons, Andrew and David, and were married until his death in 1986. Catherine also had a son from her first marriage, journalist Scott McConnell.

In the early 1960s Hayden rented one of the pilot houses of the retired ferryboat *Berkeley*, which was docked in Sausalito, California, and resided there while writing his autobiography *Wanderer*, which was published in 1963.

In the 1970s, after his appearance in *The Godfather*, he appeared several times on NBC's *Tomorrow Show* with Tom Snyder where he talked about his career resurgence and how it had funded his travels and adventures around the world. Hayden bought a canal barge in the Netherlands in 1969, eventually moving it to the heart of Paris and living on it part of the time. He also shared a home in Connecticut with his family, and had an apartment in Sausalito.

Sterling Hayden died of prostate cancer in Sausalito, California in 1986 at the age of seventy.

LLOYD HAYNES
Pete Dixon

Samuel Lloyd Haynes (September 19, 1934 - December 31, 1986) was an actor and television writer who was born in South Bend, Indiana. He was best known for his portrayal of history teacher 'Pete Dixon' on the Emmy Award-winning television series *Room 222*, and is known to 'Trekkies' for playing Lieutenant Alden in *Where No Man Has Gone Before*, the second pilot episode of the original Star Trek series.

Room 222, which was shown on ABC from September of 1969 to January of 1974 and is still in syndication, was one of the first television programs to concentrate in a compassionate and fairly realistic manner on problems that affected urban youth such as drugs, dropping out, and racial prejudice. A popular and critical success, the series won many awards from educational and civil-rights groups.

95

Haynes studied acting at the Film Industries Workshop and Actors West in Los Angeles, and served in the Marine Corps from 1952 to 1964 and. After that he became a public-affairs officer in the Naval Reserve, attaining the rank of Commander.

He also appeared in several films, including *Ice Station Zebra* with Rock Hudson and *Good Guys Wear Black* with Chuck Norris. More recently, he played the role of Mayor Ken Morgan in the daytime soap opera *General Hospital*.

Lloyd Haynes died of lung cancer in Coronado, California on December 31, 1986 at the age of fifty-two.

LOUIS HAYWARD
The Saint

Louis Charles Hayward was born on March 19, 1909 in Johannesburg, South Africa and was sent to England to be educated there and on the European Continent. With his career goals unclear Hayward spent a short time managing a London nightclub, but he displayed a talent for acting and was quickly tapped by playwright Noel Coward as his patron. Matinee idol-handsome, Hayward developed acting skills as a co-star in the London staging of several Broadway plays, among them *Dracula* and *Another Language*. He began his film career in the British *Self Made Lady* in 1932, which was followed by five UK films through the following year.

Hayward came to New York and Broadway in 1935 to star in *Point Verlaine*. It was his only Broadway venture, but it brought him a Hollywood contract. His first American film role was in 1935's *The Flame Within*. After several

97

supporting roles into 1936, he got his real break starring in the extended romantic prologue of Warner Brothers' *Anthony Adverse*. On the strength of his role as dashing officer Denis Moore, Hayward became a romantic leading man - and a swashbuckler at that.

Through the remainder of the 1930s he would have ample opportunities to vary that class of character, starting with some early B-tier efforts. Along with his good looks Hayward had an airy delivery of speech which worked as hero, and rogue, or occasional suave villain.

The familiar British Simon Templar character was brought to the screen by Hayward in *The Saint in New York* in 1938 to cap his B picture career, but he was destined for plenty of sword-point adventure. The stylish *The Man in the Iron Mask*, the third volume in the Dumas musketeer trilogy, gave him the opportunity to play the good and evil royal twins with impressive flair, but the swashbuckling efforts did not pan out as they had for Errol Flynn. *The Son of Monte Cristo* was a *Prisoner of Zenda* look-alike that fell flat. Another bad break was his 1941 casting with a pivotal role in Orson Welles' *The Magnificent Ambersons* - only to be edited out.

Then, with the onset of World War II, Hayward had a respite from the vagaries of Hollywood luck. While a Captain in the Marine Corps (of the photographic section) he and his unit filmed the Battle of Tarawa. It was the first time in the history of amphibious warfare that photographers had landed to take a beachhead with the initial assault waves. The battle was one of the bloodiest in Marine history - three days of fighting cost the Marines nearly 3,000 casualties, and over 4,500 Japanese were killed. The carnage Captain Hayward saw would lead to depression and a complete physical collapse. For his photographic services with the

Marines during the filming of the Battle of Tarawa, Hayward received a Bronze Star for courage under fire.

Overcoming the psychological stress of his experiences, he returned to the lights after the war. Already with a few mysteries under his belt, he was cast - perhaps not surprisingly - as twins in the hit 1945 Agatha Christie thriller *And Then There Were None*. Thereafter the mix of romantic parts included yet another *Monte Cristo*, the Robin Hood-like Robert Louis Stevenson adventure *The Black Arrow*, and a succession of pirate parts - particularly (and unfortunately) two sequels playing *Captain Blood*. There was also yet another twin sequel, this time a twist of the Jekyll/Hide story but with the doctor's twin sons. There was also one more outing for an iron mask vehicle, this time with twin royal sisters, with Hayward as a mature D'Artagnan. Amid all this blandness - and 'seeing double' - Hayward had developed the good business sense to save his movie career. He was one of the first to incorporate the one-percentage-of-profits deal for both theatrical and TV releases of his post-1949 films, thus ensuring a comfortable lifelong income.

Although he continued to make movies Hayward also ventured into TV, with ten *American Playhouse* theater productions, as well as some production investments of his own. In 1954 he produced and starred in the thirty-nine week TV series *The Lone Wolf* after buying exclusive rights to several of Louis Joseph Vance's original *Lone Wolf* stories.

Hayward was married three times; to Ida Lupino (1938 – 1945), Peggy Morrow Field (1946 – 1950), and June Hanson, with whom he had a son named Dana, from 1953 until his death.

Louis Hayward died of lung cancer and renal failure on February 21, 1985 in Palm Springs, California at the age of seventy-five.

GEORGE ROY HILL
My Brother's Keeper

George Roy Hill, the versatile director whose Hollywood movies included *Butch Cassidy and the Sundance Kid* and *The Sting*, was born on December 20, 1921 in Minneapolis to George R. and Helen Frances Owens Hill.

A Marine pilot in World War II and the Korean War, an actor, a Yale graduate, and a devotee of history and Bach, Hill combined scholarship and military training in his approach to his work, achieving success as a director on Broadway, in television, and in films.

His career was often characterized by a nostalgia manifest not only in his subjects - the roaring 20's, the Depression, World War I pilots, and the sinking of the Titanic - but also in his affection for the art of straightforward storytelling.

"Just as I play nothing but Bach for pleasure, so do I read nothing but history for pleasure," Hill said in a 1975 interview in *The New York Times Magazine*. "I like to be able to sit back and pick out the most fascinating facets of an era. You have a better perspective. In the present, you get too caught up in the heat of the emotions of the moment."

The interview was a rarity. In contrast to some of today's filmmakers, eager to hawk their wares from morning till late night on television, Hill was notably inaccessible. Some Hollywood figures thought him shy. Others speculated that his reticence was rooted in his reluctance to add the costs of publicity to a film's budget. He had no interest in hiring press agents or appearing on talk shows.

"The world is slow to realize that George Roy Hill not only is a vastly talented storyteller on the screen... but also cosmically cheap," Robert Redford, a close friend who starred in Hill's greatest hits, once said.

The gangly, boyish-looking Hill belonged to a generation of directors who made their mark in the so-called golden age of television in the 1950's. Like John Frankenheimer, Arthur Penn and Sidney Lumet, he gravitated to movies as the networks lost interest in serious drama, and Hollywood held out the promise of freedom from the hectic, stressful pace of television.

1969's *Butch Cassidy* starred Paul Newman and Robert Redford as good-natured bank robbers in the waning days of the Old West, and in 1973 they reunited in *The Sting* as small-time con men who pull off a big-time swindle in Depression-era Chicago. After the movies were released Hill had, for a while, the distinction of being the sole director in history to have made two of the top-ten moneymaking films.

Butch Cassidy and the Sundance Kid won Oscars for original screenplay, original score (Burt Bacharach), best song, and cinematography. Besides Academy Awards for best film and best director, *The Sting* won five other Oscars, including those for best adapted screenplay and best score (by Marvin Hamlisch, who adapted the ragtime music of Scott Joplin).

Hill's other films ranged from the Tennessee Williams comedy *Period of Adjustment* in 1962 (his Hollywood debut) to varied fare like *The World of Henry Orient*, an adaptation of the James A. Michener best seller *Hawaii*, the flapper-era musical *Thoroughly Modern Millie*, Kurt Vonnegut's dark, surreal World War II novel *Slaughterhouse-Five*, *The Great Waldo Pepper*, which was about barnstorming World War I pilots, the raucous hockey comedy *Slap Shot*, and adaptations of John Irving's novel *The World According to Garp* and John le Carré's *Little Drummer Girl.*

Of all his films *The Great Waldo Pepper*, which also starred Redford, may have been closest to Hill's heart since he had developed an affinity for music and aviation at an early age. After school he liked to visit the airport, and his hobby was to memorize the records of World War I aces. He idolized Speed Holman, a pilot "who used to make his approach to the spectators at state fairs while flying past the grandstand upside down." Naturally Hill learned to fly, and he once owned an open-cockpit Waco biplane that was built in 1930.

Hill studied music at Yale and graduated in 1943 with a Bachelor of Arts degree. He then piloted Marine transports in the South Pacific during World War II, and after the war studied music and literature under the G.I. Bill at Trinity College in Dublin.

Needing money, Hill auditioned for the Irish actor Cyril Cusack's company and made his theatrical debut in 1948 in a walk-on role in George Bernard Shaw's *Devil's Disciple* at the Gaiety Theater in Dublin. He then returned to the United States, acted Off-Broadway, and toured with Margaret Webster's Shakespeare Repertory Company, where he met Louisa Horton - whom he married on April 7, 1951. They were later divorced.

Hill was recalled to service at the Marine Corps jet flight training center in Cherry Point, North Carolina during the Korean War. While there he had to be talked down by a ground controller at the Atlanta airport one night, an incident that led to his writing *My Brother's Keeper*, a television play presented in 1953 by the *Kraft Television Theater* with Hill in the cast.

He also wrote, produced and directed television dramas like the Emmy-winning *A Night to Remember* about the sinking of the Titanic, *The Helen Morgan Story*, a biography of the torch singer, and *Judgment at Nuremberg.*

Like many filmmakers, Hill was never happy with reviewers. On the day in 1975 when Universal Studios gave him a deal granting total autonomy to do fifteen productions in the next five years, he could not forget something critic Pauline Kael had written.

"What about that Pauline Kael accusing me of emphasizing male relationships with Redford and Newman?" he said. "What am I supposed to do, stop the action in an action picture just to drag some women in?"

George Roy Hill died of complications from Parkinson's disease in his Manhattan apartment on December 27, 2002 at the age of eighty-one.

ERNIE HUDSON
Who You Gonna Call?

Ernest Lee "Ernie" Hudson is an actor who was born December 17, 1945 in Benton Harbor, Michigan. Hudson never knew his father and his mother, Maggie Donald, died of tuberculosis when he was two months old. He was subsequently raised by his maternal grandmother, Arrana Donald, and had a half-brother named Lewis Hudson. He is arguably best known for his role as Winston Zeddemore in the *Ghostbusters* film series.

Hudson joined the Marine Corps after graduating from high school, but was discharged less than three months later due to asthma. He then moved to Detroit, where he became the resident playwright at Concept East, the oldest Black Theatre company in the country. In addition he enrolled at Wayne State University to further develop his writing and acting skills, and found time to establish the Actors' Ensemble Theatre where he and other talented young black writers directed and appeared in their own works before enrolling and subsequently graduating from Yale School of Drama.

While performing with the school's repertory company he was asked to appear in the Los Angeles production of Lonne Elder III's musical *Daddy Goodness*, which led to his meeting Gordon Parks, who subsequently gave Hudson the co-starring role in his first feature film, *Leadbelly*. Unfortunately all that followed *Leadbelly* was a year of "bit parts and some harsh lessons about Hollywood," which led Hudson to enroll in another academic program at the University of Minnesota that would lead to a Ph.D. Through this experience he learned another vital lesson, saying "There are those who spend their lives studying it, and those who spend their lives doing it." Hudson definitely wanted to be in the second group. Keeping in mind this self-revelation, he accepted the starring role of Jack Jefferson in the Minneapolis Theatre-In-The-Round's production of *The Great White Hope*, a role he put "everything he had into," including shaving his head. A series of starring and guest roles followed on such television shows as *Fantasy Island, The Incredible Hulk, Little House on the Prairie, Diff'rent Strokes, Taxi, One Day at a Time, Gimme a Break!, The A-Team* and *Webster*, as well as a co-starring role in the TV mini-series *Roots: The Next Generations*. Other feature film credits include *The Jazz Singer, The Main Event* and, of course, the mega-hit *Ghostbusters*.

Off the screen Hudson as been a Reserve Deputy Sheriff in the San Bernardino County, California Sheriff's office for fourteen years (as of 2003), joining in 1994 after studying for three acting roles with the police.

DON IMUS
Imus in the Morning

John Donald "Don" Imus, Jr. is a radio host, humorist, writer, and philanthropist who was born on July 23, 1940 in Riverside, California. Although born in California, he was raised on a sprawling cattle ranch called *The Willows* near Kingman, Arizona. His nationally-syndicated talk show, *Imus in the Morning,* airs throughout the United States on Citadel Media.

Many people are surprised to hear Imus served as a bugler in the Marine Corps from 1957 to 1960, but he never fails to mention the Marine Corps' birthday on his radio show each November 10.

In 1999 Imus and his wife founded the *Imus Ranch*, a working 4,000-acre cattle ranch near Ribera, New Mexico, fifty miles southeast of Santa Fe, for children with cancer as well as siblings of SIDS victims. Between Memorial Day and Labor Day each year the Imus family volunteers their time at the Imus Ranch, and Don Imus continues his broadcasts from a studio there while the rest of his cast broadcasts from New York. In 2000 Imus suffered serious injuries after a fall from a horse while at the ranch, and ended up broadcasting several shows from a hospital.

Imus has also been instrumental in raising over sixty million dollars for the *Center for the Intrepid,* a Texas rehabilitation facility for soldiers wounded in the Iraq War. Considered to be the largest technological center of its kind in the country, it is designed to help treat disabled veterans with their transition back into the community.

More recently Imus took on the cause of the living conditions at the Walter Reed Army Medical Center, since he often visits wounded vets at the hospital and is a morale booster for these heroes. Imus' reporting preceded Army resignations, including that of Lieutenant General Kevin Kiley, then Army Surgeon General. Imus had earlier criticized Kiley's personal fitness for military duty and dedication to wounded soldiers.

In 1979 Imus divorced his first wife, Harriet, and he subsequently married his second wife, Deirdre Coleman, on December 17, 1994. He has two stepdaughters that he adopted during his first marriage, two daughters from that marriage, and one son from his current marriage. As a point of interest, both Don and Deirdre Imus are vegetarians.

Imus won three Marconi Awards, was named one of the twenty-five Most Influential People in America by *Time* magazine in 1997, and was inducted into the Radio Hall of Fame in 1989.

On March 16, 2009 Don Imus announced on his radio show that he has been diagnosed with stage two prostate cancer.

BOB KEESHAN
Captain Kangaroo

"My Marine Corps experience has served me well. Everyone has heard about the pride of the Marines... we were given a very positive feeling about our capabilities. A very high esteem, something more parents should give their children." – *Bob Keeshan*

Robert James Keeshan (June 27, 1927 - January 23, 2004) was a television producer and actor. He is most famous as the title character of the children's television program *Captain Kangaroo*, which became an icon for millions of baby boomers during its thirty-year run from 1955-1984. Keeshan also played the original "Clarabell the Clown" on the *Howdy Doody* television program.

Keeshan was born in Lynbrook, New York and attended Fordham University after serving in the Marine Corps reserve during World War II. A persistent urban myth is that actor Lee Marvin appeared on *The Tonight Show* and said he had served in the Marine Corps fighting alongside Keeshan at the Battle of Iwo Jima. Marvin never told that story, never served on Iwo Jima (having been invalided out after the battle of Saipan months earlier), and Keeshan never saw combat or overseas duty, having enlisted after turning eighteen just before the end of the war.

After World War II network television programs for children were new. On *Howdy Doody*, an early show which premiered in 1947, Keeshan played "Clarabell the Clown," a silent Auguste clown who mainly communicated by honking horns attached to a belt around his waist (one of the horns meant "yes," and the other meant "no"). Clarabell often spritzed Buffalo Bob Smith with a seltzer bottle and played practical jokes. He gave up the role in 1952, and was replaced by another actor.

In August 1953 Bob Keeshan was back on the air doing a new children's show, *Time for Fun*, playing the role of Corny the Clown - a clown who spoke. Later that same year, in addition to *Time for Fun,* Bob began *Tinker's Workshop*, a program aimed at preschoolers, where Bob played the grandfatherly Tinker.

Developing the ideas from *Tinker's Workshop*, Keeshan and long-time friend Jack Miller submitted the concept of *Captain Kangaroo* to the CBS network, which was searching for innovative new approaches to children's television programming. CBS approved the new show, and Keeshan starred as the title character of Captain Kangaroo when it premiered on October 3, 1955. Keeshan described his character as based on "the warm relationship between grandparents and children." The show was a great success, and he served as host for almost three decades. Frequently recurring characters included Mr. Green Jeans (played by fellow Marine Hugh "Lumpy" Brannum), and puppets such as "Bunny Rabbit" and "Mr. Moose."

The *New York Times* commented, "Captain Kangaroo, a round-faced, pleasant, mustachioed man possessed of an unshakable calm ... was one of the most enduring characters television ever produced."

After *Captain Kangaroo* ended Keeshan hosted 1985's *CBS Storybreak*, which featured animated versions of children's literature. Keeshan appeared in framing sequences for the animated stories, showcasing the book versions and suggesting similar books for the viewers to seek out. In 1987 Keeshan founded Corporate Family Solutions with former Tennessee Republican Governor Lamar Alexander. The company provided day-care programs to businesses.

Keeshan lived in Babylon Village on Long Island before moving to spend the last fourteen years of his life in Vermont, where he became a children's advocate as well as an author. His memoirs, entitled *Good Morning, Captain*, were published in 1995 by Fairview Press. He was a strong advocate against video game violence, and took part in the congressional hearings in 1993.

Bob Keeshan died of natural causes in Windsor, Vermont on January 23, 2004 at the age of seventy-six and was interred at Saint Joseph's Cemetery in Babylon, New York. He is survived by three children. His wife of forty years, Anne Jeanne Laurie Keeshan, had died in 1990. Bob Keeshan's grandson, Britton Keeshan, became the youngest person at the time to climb the Seven Summits when he summited Mount Everest in May 2004. He did so carrying photos of his grandfather, and buried a photo of the two of them at the summit of Everest.

HARVEY KEITEL
Mister Wolf

"(My Marine Corps service) was the first time I had a real sense of pride about myself, a sense of belonging to a group that's special. To this day, I'm proud of being a Marine." *–Harvey Keitel*

Harvey Keitel was born May 13, 1939 in the New York City borough of Brooklyn, the son of Miriam and Harry Keitel, Jewish immigrants from Romania and Poland. His parents owned and ran a luncheonette, and his father also worked as a hat maker.

Keitel grew up in the Brighton Beach section of Brooklyn with his sister Renee and brother Jerry and attended Abraham Lincoln High School. At the age of sixteen he decided to join the Marine Corps, a decision that took him to Beirut, Lebanon as part of Golf Company, 2nd Battalion, 8th Marine Regiment. After his return to the United States he was a court reporter until beginning his acting career.

Keitel studied under both Stella Adler and Lee Strasberg, eventually landing roles in some off-Broadway productions. During this time he met struggling filmmaker Martin Scorsese and gained a part in Scorsese's student production *Who's That Knocking at My Door*. Since then Scorsese and Keitel have worked together on numerous projects. Keitel had the starring role in Scorsese's *Mean Streets,* but this proved to be Robert De Niro's breakthrough film. He later appeared with De Niro in *Taxi Driver*, playing the role of a pimp.

Originally Keitel was to have played the role of Captain Willard in Francis Ford Coppola's *Apocalypse Now*, but was fired early in the production and replaced by Martin Sheen. After that it was many years before he would be able to get anything other than minor roles. At the end of the 1970s Keitel was mostly working in European films for directors such as Ridley Scott, usually in sinister character parts.

Throughout the 1980s Keitel continued to find plenty of work on both stage and screen, but was usually in the stereotypical role of a thug. This role reached its zenith when he starred in Quentin Tarantino's *Reservoir Dogs* in 1992, where his performance as 'Mr. White' re-launched his semi-slumping career. Ridley Scott also helped Keitel by casting him as the sympathetic policeman in *Thelma and Louise* in 1991. That same year he landed a role in *Bugsy*, for which he received an Academy Award nomination for Best Supporting Actor. Since then Keitel has chosen his roles with care, seeking to change his image and show off a broader acting range. One of those roles was the title character in *Bad Lieutenant*, about a self-loathing police lieutenant trying to redeem himself. His decision to co-star in Jane Campion's *The Piano* marks the approximate beginning of this phase of Keitel's career.

One of his best known roles was as efficient clean-up expert Winston Wolf in Quentin Tarantino's *Pulp Fiction*. In 1996 he landed a major role in Quentin Tarantino and Robert Rodriguez's film *From Dusk Till Dawn*, and in 1997 he starred in the crime drama *Cop Land*, which also starred Sylvester Stallone, Ray Liotta, and De Niro. Later roles include the fatherly Satan in *Little Nicky*, a wise Navy man in *U-571*, and diligent FBI agent Sadusky in *National Treasure*. Keitel's pride in being a Marine was evident in the latter, as he wore a red necktie emblazoned with gold Eagle, Globe and Anchors throughout the movie.

Unlike many American male actors who never appear nude on film or only do so once, Keitel has done it in several films - including full frontal nudity in *Bad Lieutenant* and *The Piano*.

In 2002 Keitel was honored with the Stanislavsky Award for outstanding achievement in a career of acting and devotion to the principles of Stanislavsky's school in a ceremony at the Moscow International Film Festival.

Keitel was formerly in a long-term relationship with actress Lorraine Bracco, and married actress Daphna Kastner in 2001. He is the father of three children: daughter Stella (from his relationship with Bracco), son Hudson (from a relationship with Lisa Karmazin), and son Roman (with Kastner).

BRIAN KEITH
Uncle Bill

Brian Keith (November 14, 1921 - June 24, 1997), born Robert Keith Richey Jr. in Bayonne, New Jersey to actor Robert Keith and stage actress Helena Shipman, was a stage, film and television actor. He made his acting debut in 1924 in the silent film *Pied Piper Malone* at the age of three. From 1927 through 1929 Keith's stepmother was Peg Entwistle, a failed actress who committed suicide by jumping from the letter "H" in the famous Hollywood sign.

After high school in East Rockaway, New York Keith joined the Marine Corps, serving during World War II from 1942 to 1945 as an air gunner and receiving an Air Medal.

After the war he became a stage actor, branching out into films and then television. A strong and capable actor, Keith spent many years playing second leads and gruff sidekicks, and won much acclaim for his starring role in fellow Marine Sam Peckinpah's short-lived *The Westerner* in 1960. His biggest break, however, came in 1966 when he landed the role of 'Uncle Bill' Davis on the popular television situation comedy *Family Affair*, a role that earned him three Emmy nominations for Best Actor and made him a household name. When CBS requested that he pose for Christmas publicity shots connected with *Family Affair*, Keith refused on the basis that it was exploitative of the holiday.

He is fondly remembered for his role as the father of twins in the 1961 film *The Parent Trap,* costarring Hayley Mills and Maureen O'Hara. His performance as Theodore Roosevelt in 1975's *The Wind and the Lion* is also particularly well-remembered, and is considered among the best portrayals of an American president on film. Ironically the film revolved around his decision to land Marines in Morocco, and the scenes depicting them fighting in dress blues are a favorite amongst today's leathernecks.

Keith went on to star in such television series as *The Brian Keith Show, Heartland,* and *Hardcastle and McCormick.* He also starred as Stephen 'The Fox' Halliday in the six-part television series *The Zoo Gang*, which was about a group of former World War II underground freedom fighters.

Keith spoke fluent Russian, which led to his casting as a Russian in two roles - the Soviet Premier in *World War III* with Rock Hudson, and as a Soviet scientist in *Meteor* with Natalie Wood. Ironically, in *The Russians Are Coming, The Russians Are Coming* he played the unexcitable police chief of an island where a Soviet submarine runs aground and his character had to have Russian translated to him by Alan Arkin's character.

Two films of particular interest to Marines were1987's *Death Before Dishonor*, in which Keith's character, Colonel Halloran, is kidnapped and tortured by Islamic terrorists, and the aforementioned *Wind and the Lion* in which President

Theodore Roosevelt (Keith) dispatches Marines to rescue a kidnapped American woman in 1904 Morocco.

In his last film Keith played President William McKinley in 1997's *Rough Riders*. Director John Milius dedicated *Rough Riders* to "Brian Keith... Actor, Marine, Raconteur."

Keith married three times, first to Frances Helm, then in 1955 to Judith Landon, and finally in 1970 to Hawaiian actress Victoria Young (née Leialoha), who later appeared on *The Brian Keith Show* (1972-1974) as Nurse Puni. Keith fathered four children, but also adopted three others with Judith Landon. Daisy Keith, one of his children with Victoria Young, became an actress and appeared with her father in the short-lived series *Heartland* in 1989.

During the later part of his life Keith suffered from emphysema and lung cancer, despite having quit smoking ten years earlier (he had posed for Camel cigarettes in an endorsement campaign in 1955).

Brian Keith was found dead of a self-inflicted gunshot wound on June 24, 1997, two months after his daughter Daisy had committed suicide. He is buried next to her at Westwood Village Memorial Park Cemetery in Los Angeles, California. On June 26, 2008 the Hollywood Walk of Fame installed a star in Brian Keith's honor.

DAN LAURIA
Jack Arnold

Daniel Joseph "Dan" Lauria is a television and film actor who was born on April 12, 1947 in Brooklyn, New York, the son of Carmella (née Luongo) and Joseph J. Lauria. An Italian-American, Lauria also lived in Lindenhurst, New York for a period of time. He graduated from Lindenhurst High School in 1965 as a Varsity Football player, and briefly taught physical education there. A Vietnam War veteran, Lauria served as an officer in the Marine Corps in the early 1970s and attained the rank of Captain, just as his character 'Jack Arnold' did during the Korean War.

Lauria got his start in acting while attending Southern Connecticut State University in New Haven, Connecticut and is best known for his portrayal of Jack Arnold on the TV series *The Wonder Years* from 1988 to 1993. He also played James Webb in the 1998 TV miniseries *From the Earth to the Moon* and *Independence Day* in 1996. Recently he has

117

appeared in a War Veterans public service announcement and as Police Commissioner Eustace Dolan on *The Spirit.* He also appeared onstage in New York in the summer of 2006 in an Off-Broadway production of William Mastrosimone's *A Stone Carver*, and also had TV roles on *Army Wives, The Mentalist* and *Criminal Minds.*

JIM LEHRER
Dean of Moderators

"While in the Marines I had a chance to test myself, physically and emotionally and spiritually, in important lasting ways." - *Jim Lehrer*

James Charles Lehrer is a journalist and news anchor who was born on May 19, 1934 in Wichita, Kansas, the son of Harry Frederick Lehrer, a bus station manager, and Lois Catherine (née Chapman), a bank clerk. He is best known for *The News Hour with Jim Lehrer* on PBS, and for his role as a frequent debate moderator during elections. Lehrer is also an author of non-fiction and fiction books, drawing from his experiences and interest in history and politics.

Lehrer attended middle school in Beaumont, Texas and graduated from Thomas Jefferson High School in San Antonio, Texas where he was one of the three sports editors at the *Jefferson Declaration*. He also graduated from Victoria College in Texas and the Missouri School of

Journalism at the University of Missouri. He then joined the Marine Corps and attained the rank of captain before mustering out in 1959, and attributes his service and travels with helping him to look past himself and feel a connection to the world that he would not have otherwise experienced.

Lehrer began his career in journalism at *The Dallas Morning News* and the *Dallas Times-Herald*, where he covered the assassination of John F. Kennedy in 1963, and from 1970 to 1973 he anchored the local single-story news show *Newsroom* on KERA-TV, the local PBS affiliate in Dallas. Lehrer began working with the PBS network in 1973, and in 1975 developed and co-anchored *The MacNeil/Lehrer Report* with Robert MacNeil. The show was later renamed *The MacNeil/Lehrer News Hour*, and then after MacNeil's departure in 1995 was ultimately named *The News Hour with Jim Lehrer.*

Nicknamed the "Dean of Moderators" by fellow Marine Bernard Shaw of CNN, Lehrer has moderated eleven presidential candidate debates, with the most recent being the presidential debate between senators Barack Obama and John McCain on September 26, 2008.

Lehrer is a bus enthusiast, and is a supporter of the Pacific Bus Museum in Williams, California and the Museum of Bus Transportation in Hershey, Pennsylvania because his father was a bus driver and also briefly operated a bus company - and he himself once worked as a Trailways ticket agent in Victoria, Texas while a college student in the 1950s.

The lengthy list of awards and honors earned by Lehrer include the George Foster Peabody Award National Humanities Medal, American Academy of Arts and Sciences Fellow, University of Missouri School of Journalism's Medal of Honor, William Allen White Foundation Award for Journalistic Merit, Fred Friendly First Amendment Award,

National Academy of Television Arts and Sciences Silver Circle, Television Hall of Fame, and an Emmy.

Jim Lehrer is married to novelist Kate Lehrer, and has three children and six grandchildren.

JOE LISI
Lieutenant Swersky

"I firmly believe that becoming a Marine was the defining experience of my life. Any success I have achieved, I feel I owe in no small measure to the values and principles I learned in the Marine Corps." – *Joe Lisi*

Joe Lisi, also credited as Joe Lissi, is a television actor who was born on September 9, 1950 in New York City. He is best known for appearing as Lieutenant Swersky on the NBC television show *Third Watch* from 2000 to 2005. He has also appeared as parole officer Craig Lennon on the NBC television show *Law & Order: Special Victims Unit*, and as Dick Barone on *The Sopranos.*

Lisi spent twenty-four years in the New York Police Department (NYPD), retiring at the rank of captain. In 1969, while already employed by the police department, he enlisted in the Marine Corps Reserve and was eventually Honorably Discharged as a corporal.

LEE MARVIN
The Dirty Dozen

"I was a PFC in the Marine Corps, so when I started playing officers (in the movies) I had a good opinion as to how they should be portrayed – from the bias of an enlisted man's viewpoint." – *Lee Marvin*

Lee Marvin (February 19, 1924 - August 29, 1987) was a film actor known for his gravelly voice, white hair and 6'2" stature. Marvin at first did supporting roles, mostly villains, soldiers and other hard-boiled characters, but after winning a Best Actor Oscar for his dual roles in *Cat Ballou* he landed more heroic and sympathetic leading roles.

Marvin was born in New York City, the son of Lamont Waltman Marvin, an advertising executive and head of the New York and New England Apple Institute, and his wife Courtenay Washington Davidge, a fashion writer and beauty consultant. His father was a direct descendant of Matthew

Marvin, Sr., who emigrated from England in 1635 and helped found Hartford, Connecticut.

Marvin studied violin when he was young, and as a teenager "spent weekends and spare time hunting deer, puma, wild turkey and bobwhite in the wilds of the then-uncharted Everglades." He attended St. Leo Preparatory College in St. Leo, Florida after being expelled from several schools for bad behavior.

Marvin left school to join the 4th Marine Division, serving as a sniper. He was wounded in action during the Battle of Saipan during WWII eight months prior to the Battle of Iwo Jima, and most of the members of his platoon were killed. This had a significant effect on Marvin for the rest of his life. He was awarded the Purple Heart medal and was given a medical discharge with the rank of Private First Class. Contrary to rumors, Marvin did *not* serve with Bob Keeshan during World War II.

While working as a plumber's assistant at a local community theater in upstate New York Marvin was asked to replace an actor who had fallen ill during rehearsals. He then began an amateur off-Broadway acting career in New York City and served as an understudy in Broadway productions.

In 1950 Marvin moved to Hollywood where he quickly found work in supporting roles, and from the beginning was cast in various Western or war films. As a decorated combat veteran Marvin was a natural in war dramas, where he frequently assisted the director and other actors in realistically portraying infantry movement, arranging costumes, and even adjusting war surplus military prop firearms. His debut was in 1951's *You're in the Navy Now*, and in 1952 he appeared in several films, including Don Siegel's *Duel at Silver Creek, Hangman's Knot*, and the war drama *Eight Iron Men.* He then played Gloria Grahame's vicious boyfriend in Fritz Lang's *The Big Heat.* Marvin then had a small but memorable role in *The Wild One* opposite Marlon Brando (Marvin's motorcycle gang in the film was called "The Beetles"), followed by *Seminole* and *Gun Fury.* He was again praised for his role as Hector the small town hood in *Bad Day at Black Rock* with Spencer Tracy.

During the mid-1950s Marvin gradually began playing more substantial roles. He starred in *Attack* and *The Missouri Traveler*, but it took over one hundred episodes as Chicago cop Frank Ballinger in the successful 1957-1960 television series *M Squad* to actually give him name recognition. One critic described the show as "a hyped-up, violent Dragnet... with a tough-as-nails Marvin" playing a police lieutenant.

In the 1960s Marvin was given prominent co-starring roles in films such as *The Comancheros, The Man Who Shot Liberty Valance* (Marvin played Liberty Valance) and *Donovan's Reef*, all with John Wayne. He also guest-starred on *Combat!,* and *The Twilight Zone* during that period.

Thanks to director Don Siegel, Marvin appeared in the groundbreaking *The Killers,* playing an organized, no-nonsense, efficient, businesslike professional assassin whose character was copied to a great degree by Samuel L. Jackson in the 1994 Quentin Tarantino film *Pulp Fiction. The Killers* was also the first movie in which Marvin received top billing and the only time Ronald Reagan played a villain.

Marvin won the 1965 Academy Award for Best Actor for his comic role in the offbeat western *Cat Ballou* starring Jane Fonda. Following roles in *The Professionals* and the hugely successful *The Dirty Dozen*, he was given complete control over his next film. In *Point Blank*, an influential film with director John Boorman, he portrayed a hard-nosed criminal bent on revenge. In that film Marvin, who had selected

Boorman himself for the director's slot, had a central role in the film's development, plot line, and staging. In 1968 he also appeared in another Boorman film as a Marine in the critically acclaimed but commercially unsuccessful *Hell in the Pacific*, co-starring famed Japanese actor Toshirō Mifune. He even had a hit song with *Wand'rin' Star* from the 1969 western musical *Paint Your Wagon*. By this time he was getting paid a million dollars per film, $200,000 less than Paul Newman was making at the time. He was also ambivalent about the business, even with its financial rewards, saying "You spend the first forty years of your life trying to get in this fucking business, and the next forty years trying to get out. And then when you're making the bread, who needs it?"

Marvin had a much greater variety of roles in the 1970s and 1980s, with fewer 'bad-guy' roles than in earlier years. His 1970s films included *Monte Walsh*, *Prime Cut*, *Pocket Money*, *Emperor of the North Pole*, *The Iceman Cometh*, *The Spikes Gang*, *The Klansman*, *Shout at the Devil*, *The Great Scout* and *Avalanche Express*. Marvin was also offered the role of Quint, which eventually went to Robert Shaw, in *Jaws,* but declined.

Marvin's last big role was in Samuel Fuller's *The Big Red One* in 1980. His remaining films were *Death Hunt*, *Gorky Park*, *Dog Day*, *The Dirty Dozen: The Next Mission*, with his final appearance being in *The Delta Force* in 1986.

A father of six, Marvin was married twice. His first marriage to Betty Ebeling began in February 1951 and ended in divorce on January 5, 1967. During that time his hobbies included sport fishing off the Baja California coast and duck hunting along the Mexican border near Mexicali. He then married Pamela Feeley on October 18, 1970 and remained her husband until his death. During the 1970s Marvin resided off and on in Woodstock, New York and would make regular trips to Cairns, Australia to engage in marlin fishing.

In 1971 Marvin was sued by long-time girlfriend Michelle Triola (who called herself 'Michelle Marvin' at the time). Though the couple never married, she sought financial compensation similar to that available to spouses under California's alimony and community property laws. The result was the landmark 1976 "palimony" case, *Marvin v. Marvin*. In 1979 Marvin was ordered to pay $104,000 to Triola for "rehabilitation purposes," but her community property claim for one-half of the $3.6 million which Marvin had earned during their six years of cohabitation was denied. In August 1981, however, the California Court of Appeals reversed this decision, declaring Triola was entitled to no money whatsoever, in that the co-habitant in an unmarried cohabitative relationship has no community property claim, but merely a contract claim. Without evidence of any contract between Marvin and Triola requiring that Marvin support her should their relationship end, Triola could not recover any money.

Lee Marvin died of a coma-induced heart attack on August 29, 1987, and is interred at Arlington National Cemetery.

TIM MATHESON
Otter

Timothy Lewis Matthieson is an actor who was born in Glendale, California on December 31, 1947. Raised in the San Fernando Valley by a family far removed from the workings of the entertainment industry, Matheson immersed himself in his own world of entertainment, writing and performing mini-plays at home with friends. At a very young age he enrolled in an acting class, and immediately earned the attention of an agent. He started his career as a child actor and appeared on such classic TV shows as *Leave It to Beaver* and *My Three Sons*, and was the voice of *Johnny Quest* on that animated series.

Matheson made a smooth transition to adult actor following some terrific teen turns in the excellent comedy features *Divorce American Style* and *Yours, Mine and Ours*. Typecast as a cowboy during his four years as a contract player at Universal, he acted in NBC's *The Virginian* and *Bonanza* and later starred opposite Kurt Russell in their own short-lived Western series, *The Quest*. Matheson raised his feature profile as 'Sweet,' one of the rogue cops of the *Dirty Harry* sequel *Magnum Force*, and then gained fame for his full-bodied characterization of the fast-talking, womanizing, preppy snob Eric 'Otter' Stratton in John Landis' *National Lampoon's Animal House* in 1978. For years afterward he had to explain to disappointed fans that no, Otter was only a character he played, and he couldn't really offer them sexual advice.

Soon after appearing in *Yours, Mine and Ours* Matheson joined the Marine Corps Reserves and pursued collegiate studies while under a four-year contract with Universal. During this period he was given a role in virtually every television project Universal produced including *The Virginian*, *The Bold Ones*, *Night Gallery*, *Bonanza* and a pilot with Sally Field titled *Hitched*,

In recent times Matheson has worked on the USA series *Burn Notice* with producer Jeff Freilich, and as a result was offered an opportunity to direct and appear in *Behind Enemy Lines: Columbia.* In an interview he said, "There were a couple of things that were sort of in my favor. They liked the way I shot, the way I approach the work. The kicker too, the bonus for them was I was a former Marine, so I knew the military. I'm not a gung-ho kind of guy, but I do honor our military. If you're going to do a military movie, you've got to do it right. I think that all sort worked into my favor. The guys went through - we only had a week for them - but we gave them a hell week, a week of boot camp and military training and guns and how to fire the weapons, how to march and how a squad works. We had really cracker-jack guys. We had a real Navy Seal named Jeffrey Reeves with us and Tom Minder was our military advisor, a former Marine, so that was good."

ED McMAHON
Heeeeree's Johnny!

"My pride in the Marine Corps has never diminished."
– Ed McMahon

Edward "Ed" Leo Peter McMahon, Jr. (March 6, 1923-June 22, 2009) was a comedian, game show host, announcer, and television personality most famous for his work on television as Johnny Carson's announcer on *Who Do You Trust?* from 1957 to 1962 and *The Tonight Show* from 1962 to 1992, and as the host of the talent show *Star Search* from 1983 to 1995. He later also became well-known as the presenter of American Family Publishing sweepstakes (not to be confused with Publishers Clearing House), which arrives unannounced at the homes of winners. He subsequently made a series of *Neighborhood Watch* Public Service Announcements reprising that role in parody.

He also co-hosted the *Jerry Lewis Labor Day Telethon* and performed in numerous television commercials, most notably for Budweiser. In the 1970s and 1980s he anchored the team of NBC personalities conducting the network's coverage of the Macy's Thanksgiving Day Parade.

McMahon appeared in several films including *The Incident*, *Fun With Dick and Jane*, *Full Moon High*, *Butterfly* and a brief bit in the film version of *Bewitched* in 2005.

McMahon was born in Detroit, Michigan, the son of Eleanor (née Russell) and Edward Leo McMahon, who was a fund-raiser and entertainer. He was raised in Lowell, Massachusetts and attended Boston College and The Catholic University of America, majoring in speech and drama. At The Catholic University of America he joined the Phi Kappa Theta fraternity and graduated with a Bachelor of Arts degree in 1949.

McMahon began his career as a bingo caller in Maine when he was fifteen. Prior to working as an announcer he worked as a carnival barker for three years during his teenage years in Maine, and put himself through college as a pitchman for vegetable slicers on the Atlantic City boardwalk.

His first broadcasting job was at WLLH-AM in Lowell, and he began his television career in Philadelphia at WCAU-TV. In the 1960s he emceed the game shows *Missing Links* (when the show moved to ABC, Dick Clark replaced him), *Snap Judgment*, *Concentration* and *Who Dunnit?*

During World War II McMahon was trained as a Marine fighter pilot, and also served as a flight instructor and test pilot. He was discharged in 1946, but remained in the reserves.

After college McMahon returned to active duty and was sent to Korea in February of 1953, where he flew unarmed

O-1E Bird Dogs on eighty-five tactical air control and artillery spotting missions. He remained in the Marine reserves, retiring as a Colonel in 1966, and was then commissioned a Brigadier General in the California Air National Guard.

Several of his ancestors, including the Marquis d'Equilly, also had long and distinguished military careers. Patrice MacMahon, duc de Magenta was a Marshal of Armies in France, serving under Napoleon III. McMahon once asserted to Johnny Carson that mayonnaise was originally named 'Macmahonnaise' in honor of this ancestor, referring to him as the Comte de MacMahon. In his autobiography he said it was his father who told him of this relationship, and went on to suggest he was not certain of the truth of the story.

McMahon and Johnny Carson first worked together as announcer and host on the daytime game show *Who Do You Trust?* from 1957 to1962, and then they left to do *The Tonight Show* in 1962. For more than thirty years thereafter McMahon introduced the *Tonight Show* with his trademark drawn-out "Heeeeeeeeeeeeeeere's Johnny!" and his booming voice and constant laughter earned him the nickname the "Human Laugh Track."

The extroverted McMahon served as a counter to the notoriously shy Carson. Nonetheless, McMahon once told an interviewer that even after his many decades as an emcee he would still get "butterflies" in his stomach every time he would walk onto a stage, and would use that nervousness as a source of energy.

Ed McMahon's status as an American icon is secure, and pop culture references to his career are legion. Comedian Garry Shandling stated in interviews that the relationship between fictional talk show host Larry Sanders and his side-kick Hank Kingsley in the hit sitcom *The Larry Sanders*

Show was largely based on McMahon and Carson, Jack Nicholson's character in *The Shining* famously shouted "Heeeeeeeeeeeeeeere's Johnny!" while holding an axe and attempting to kill his wife, and musical comedy icon "Weird Al" Yankovic wrote a parody of El Debarge's hit *Who's Johnny* entitled *Here's Johnny* about Ed McMahon and his signature catchphrase which appeared on his 1987 album *Polka Party!*

In addition to his long run on the *Tonight Show* McMahon also hosted of the successful weekly syndicated series *Star Search*, which began in 1983 and helped launch the careers of numerous actors, singers, choreographers and comedians. He stayed with the show until it ended in 1995, and in 2003 made a cameo appearance on a revival of the show hosted by Arsenio Hall, who was his successor.

McMahon and Dick Clark hosted the TV series (later special broadcast) *TV Bloopers and Practical Jokes* on NBC from 1982 until 1998, when Clark decided to move the production of the series to ABC.

In 2004 he became the announcer and co-host of *Alf's Hit Talk Show* on TV Land. Ed McMahon authored two memoirs, *Here's Johnny! My Memories of Johnny Carson, The Tonight Show, and 46 Years of Friendship* and *For Laughing Out Loud.*

Ed McMahon died in Los Angeles, California on June 23, 2009 at the age of eighty-six and was interred at the Forest Lawn-Hollywood Hills Cemetery.

STEVE McQUEEN
The King of Cool

"I liked being in the Marines. They gave me discipline I could live with... sure I was pretty wild – but I had a lot of rough edges knocked off." – *Steve McQueen*

Terrence Steven "Steve" McQueen (March 24, 1930 - November 7, 1980) was a movie actor who was born in Beech Grove, Indiana, a suburban community bordering Indianapolis. Nicknamed "The King of Cool," his anti-hero persona, which he developed at the height of the Vietnam counterculture, made him one of the top box-office draws of the 1960s and 1970s. After appearing in the 1974 film *The Towering Inferno* he became the highest paid movie star in the world. Although McQueen was combative with directors and producers, his popularity put him in high demand and enabled him to command large salaries.

He was an avid racer of both motorcycles and cars. While he studied acting he supported himself partly by competing in weekend motorcycle races and bought his first motorcycle with his winnings. He is recognized for performing many of his own stunts, especially the stunt driving during the high-speed chase scene in *Bullitt*. Additionally, McQueen designed and patented a bucket seat for race cars.

His father William was a stunt pilot for a barnstorming flying circus who abandoned McQueen and his mother when McQueen was six months old, and his mother Jullian was a young, rebellious alcoholic. Unable to cope with bringing up a small child, she left him with her parents (Victor and Lillian) in Slater, Missouri, in 1933. Shortly thereafter, as the Great Depression set in, McQueen and his grandparents moved in with Lillian's brother Claude on the latter's farm in Slater.

McQueen had good memories of the time spent on his Great Uncle Claude's farm. In recalling Claude, he stated "He was a very good man, very strong, very fair. I learned a lot from him." On McQueen's fourth birthday Claude gave him a red tricycle, which McQueen later claimed started his interest in racing. At the age of eight he was taken back by his mother and lived with her and her new husband in Indianapolis. McQueen retained a special memory of leaving the farm. "The day I left the farm Uncle Claude gave me a personal going-away present - a gold pocket watch, with an inscription inside the case." The inscription read, "To Steve - who has been a son to me."

McQueen, who was dyslexic and partially deaf as a result of a childhood ear infection, did not adjust well to his new life. Within a couple of years he was running with a street gang and committing acts of petty crime. Unable to control his behavior, his mother sent him back to Slater again. A couple of years later, when McQueen was twelve, Jullian wrote to Claude asking that Steve be returned to her once again, to live in her new home in Los Angeles. Jullian, whose second marriage had ended in divorce, had married for a third time.

This move would begin an unsettled period in McQueen's life. By his own account, he and his new stepfather "locked horns immediately." McQueen recounted him as "a prime son of a bitch" who was not averse to using his fists on both Steve and his mother. As McQueen began to rebel once again, he was sent back to live with Claude a final time. At the age of fourteen he left Claude's farm without saying goodbye and joined a circus for a short time, after which he slowly drifted back to his mother and stepfather in Los Angeles and resumed his life as a gang member and petty criminal. On one occasion McQueen was caught stealing hubcaps by the police, who proceeded to hand him over to his stepfather. The latter proceeded to beat him severely, and he ended the fight by throwing McQueen down a flight of stairs. Steve looked up at his stepfather and said, "You lay your stinkin' hands on me again and I swear, I'll kill ya."

After this McQueen's stepfather convinced Jullian to sign a court order stating her son was incorrigible and remanding him to the California Junior Boys Republic in Chino Hills, California. While there McQueen slowly began to change and mature, but he was not popular with the other boys at first. "Say the boys had a chance once a month to load into a bus and go into town to see a movie. And they lost out because one guy in the bungalow didn't get his work done right. Well, you can pretty well guess they're gonna have something to say about that. I paid his dues with the other fellows quite a few times. I got my lumps, no doubt about it. The other guys in the bungalow had ways of paying you back for interfering with their well-being." Ultimately, however, McQueen decided to give Boys Republic a fair shot. He became a role model for the other boys when he was elected to the Boys Council, the group which made the rules and regulations governing the boys' lives. He would eventually leave Boys Republic at sixteen, and when he later became famous regularly returned to talk to the boys there. He also personally responded to every letter he received from the boys, and maintained a lifelong association.

After McQueen left Chino he returned to Jullian, who was now living in Greenwich Village, but left again almost immediately. He then met two sailors from the Merchant Marine and volunteered to serve on a ship bound for the Dominican Republic. Once there he abandoned his new post, and eventually made his way to Texas and drifted from job to job. He worked as a towel boy in a brothel, an oil rigger, a trinket salesman in a carnival, and a lumberjack.

In 1947 McQueen joined the Marine Corps and was quickly promoted to Private First Class and assigned to an armored unit. Initially he reverted to his prior rebelliousness, and as a result was busted to Private on seven different occasions. Once he went AWOL by failing to return after a weekend pass had expired, instead staying away with a girlfriend for two weeks until the shore patrol caught him. He responded to his captors by resisting them and as a result spent forty-one days in the brig.

After this McQueen resolved to focus his energies on self-improvement and embraced the Marines' discipline. He saved the lives of five other Marines during an Arctic exercise, pulling them from a tank before it broke through ice into the sea. He was also assigned to the honor guard responsible for guarding then-President Harry Truman's yacht. McQueen served in the Corps until 1950, when he was honorably discharged.

In 1952, with financial assistance provided by the G.I. Bill, McQueen began studying acting at Sanford Meisner's *Neighborhood Playhouse*. He also began to earn money by competing in weekend motorcycle races at Long Island City Raceway and soon purchased the first of many motorcycles, a used Harley Davidson. He became an excellent racer, and came home each weekend with about a hundred dollars in winnings.

After several roles in productions including *Peg o' My Heart, The Member of the Wedding,* and *Two Fingers of Pride*, McQueen landed his first film role in *Somebody Up There Likes Me*, which was directed by Robert Wise and starred Paul Newman. He made his Broadway debut in 1955 in the play *A Hatful of Rain*, starring Ben Gazzara. When he appeared in a two-part television presentation titled *The Defenders*, Hollywood manager Hilly Elkins (who managed McQueen's first wife, Neile) took note of him and decided that B-movies would be a good place for the young actor to make his mark. McQueen was subsequently hired to appear in the films *Never Love a Stranger, The Blob*, and *The Great St. Louis Bank Robbery*.

McQueen's first breakout role would not come in film, but on TV. Elkins successfully lobbied Vince Fennelly, producer of the Western series *Trackdown*, to have him read for the part of a bounty hunter named Josh Randall in a new pilot for a *Trackdown* companion series. The Josh Randall character, played by Robert Culp, was introduced in an episode of *Trackdown*, after which McQueen filmed the pilot episode. The pilot was approved for a new series, now titled *Wanted: Dead or Alive* which debuted on CBS in September of 1958.

McQueen would ultimately make this role his own and become a household name as a result. Randall's holster held a sawed-off Winchester rifle nicknamed the "Mare's Leg," instead of the standard six-gun carried by the typical Western character. This added to the anti-hero image of a man infused with a mixture of mystery, alienation, and detachment to make this show stand out from the typical TV Western. Ninety four episodes, filmed at Apacheland Studio from 1958 till early 1961, kept McQueen steadily employed in television.

At twenty-nine he got his most significant break when Frank Sinatra removed Sammy Davis, Jr. from the film *Never So Few*, and Davis' role went to McQueen. Sinatra saw something special in him, and ensured that the young actor got plenty of good shots and close-ups in a role that earned McQueen favorable reviews. His character, Bill Ringa, much like other characters he would come to play, brought a new kind of cool to the screen and was never more comfortable than when driving at high speed - in this case at the wheel of a jeep. John Sturges directed this film, and then used McQueen in *The Magnificent Seven* a year later and *The Great Escape* in 1963.

After *Never So Few* director Sturges cast McQueen in his next movie, promising to "give him the camera." *The Magnificent Seven*, with Yul Brynner, Robert Vaughn, Charles Bronson and James Coburn, became McQueen's first major hit and led to his withdrawal from his own successful television series, *Wanted: Dead or Alive.*

McQueen's next big film, 1963's *The Great Escape*, told the true story of an historical mass escape from a World War II POW camp. Insurance concerns prevented him from performing the film's widely noted motorcycle leap, which was instead done by his friend and fellow cycle enthusiast

Bud Ekins - who resembled McQueen from a distance. When Johnny Carson later tried to congratulate McQueen for the jump during a broadcast of *The Tonight Show* McQueen said, "It wasn't me. That was Bud Ekins."

In 1966 McQueen earned his only Academy Award nomination for his role in the film *The Sand Pebbles.* He followed with another successful film, 1968's *Bullitt*, which featured an unprecedented (and endlessly imitated) auto chase through San Francisco, with Bud Ekins again doubling for some of the more hazardous work. McQueen also appeared in the 1971 car race drama *Le Mans*, and in *The Getaway* in 1972. He played the leading role in *Junior Bonner* in 1972, and in 1973's *Papillon.*

By the time of *The Getaway* McQueen was the world's highest paid actor. After 1974's *The Towering Inferno*, co-starring with his long-time rival Paul Newman, he did not return to film until 1978 with *An Enemy of the People,* playing against type as a heavily-bearded, bespectacled doctor in an adaptation of the Henrik Ibsen play. The film was little seen. His last films were *Tom Horn* and *The Hunter*, both released in 1980.

McQueen was offered the lead role in *Breakfast at Tiffany's* but was unable to accept due to his *Wanted: Dead or Alive* contract, so the role went to fellow Marine George Peppard. He also turned down *Ocean's Eleven, Butch Cassidy and the Sundance Kid, The Driver, Apocalypse Now,* and *Dirty Harry.* He was also the first choice of director Steven Spielberg for his film *Close Encounters of the Third Kind.*

According to Spielberg in a documentary on the *Close Encounters* DVD Spielberg met him at a bar where McQueen drank beer after beer. Before leaving the bar McQueen told Spielberg he could not accept the role because he was unable to cry on film, and the role eventually went to Richard Dreyfuss.

Steve had been interested in starring in *First Blood*, but could not due to illness. He had also been offered the titular role in *The Bodyguard* when it was first proposed in 1976. He was to play the lead in *Quigley Down Under*, which was scheduled for production in 1980, but due to his illness the project was scrapped until a decade later with Tom Selleck in the starring role.

McQueen was also interested in making the film version of *Waiting for Godot,* Because during his time away from film he developed an interest in the classic playwrights. This led him to Beckett's *Godot*, but the playwright had never heard of McQueen.

Since McQueen was an avid motorcycle and racecar enthusiast he performed many of his own stunts whenever he had the opportunity. Perhaps the most memorable were the classic car chase in *Bullitt* and the motorcycle chase scene, and although the jump over the fence in *The Great Escape* was actually done by Bud Ekins McQueen did have a considerable amount of screen time riding his motorcycle. According to the commentary track on *The Great Escape* DVD it was difficult to find riders as skilled as McQueen, and at one point in the film, due to clever editing, McQueen is seen in a German uniform chasing himself on another bike.

Together with John Sturges, McQueen planned to make *Day of the Champion*, a movie about Formula One racing, but was busy with the delayed *Sand Pebbles*. They had a contract with the German *Nürburgring*, and after John Frankenheimer shot scenes there for *Grand Prix* the reels had to be turned over to Sturges. Frankenheimer was ahead of schedule anyway, and the McQueen/Sturges project was called off.

During his acting career McQueen considered becoming a professional race car driver. In the 1970 '12 Hours of Sebring' race Peter Revson and McQueen (driving with a cast on his left foot from a motorcycle accident two weeks before) won with a Porsche 908/02 in the 3-litre class and missed winning overall by a scant twenty-three seconds to Mario Andretti, Ignazio Giunti, and Nino Vaccarella in a 5-litre Ferrari 512S. The same Porsche 908 was entered by his production company *Solar Productions* as a camera car for *Le Mans* in the 1970 *24 Hours of Le Mans* later that year. McQueen wanted to drive a Porsche 917 with Jackie Stewart in that race, but his film backers threatened to pull their support if he drove. Faced with the choice of driving for twenty-five hours in the race or driving the entire summer making the film McQueen opted to do the latter, however the film was a box office flop that almost ruined his career. McQueen admitted that he almost died while filming the movie, but nonetheless *Le Mans* is considered to be the most historically realistic, accurate, and dramatic representation of one of the most famous periods in the history of the race, as well as being considered one of the greatest auto racing movies of all time.

McQueen also competed in off-road motorcycle racing. His first off-road motorcycle was a Triumph 500cc that he purchased from friend and stunt man Bud Ekins. He raced in many of the top off-road races on the West Coast during the '60s and early 1970s, including the Baja 1000, the Mint 400 and the Elsinore Grand Prix. In 1964 he represented the United States in the International Six Days Trial, a form of off-road motorcycling Olympics. He was inducted in the Off-road Motorsports Hall of Fame in 1978. In 1971, *Solar*

Productions funded the now-classic motorcycle documentary *On Any Sunday*, in which McQueen himself is featured, along with racing legends Mert Lawwill and Malcolm Smith. Also in 1971, McQueen was on the cover of *Sports Illustrated* magazine riding a Husqvarna dirt bike.

McQueen owned several exotic sports cars, but to his dismay he was never able to own the legendary Ford Mustang GT that he drove in *Bullitt*, which featured a highly-modified drivetrain which suited McQueen's driving style. There were two cars used for filming, and according to the October 2006 issue of *Motor Trend Classic* in its cover story on the film one of the cars was so badly damaged during filming that it was judged to be unrepairable and scrapped. The second car still exists, but the owner consistently refused to sell it at any price and planned a "minimal restoration" to make the car roadworthy - while still retaining the original patina.

Towards the end of his life McQueen became a Christian, due in part to the influence of his flying instructor Sammy Mason and wife Barbara Minty. He regularly attended his local church, and was visited by evangelist Billy Graham shortly before he died. In an interview recorded shortly before his death, and as chronicled in Chopher Sandford's biography of the star, McQueen publicly lamented the fact that he would never have time to share his faith.

After discovering a mutual interest in racing, James Garner and McQueen became good friends. Garner lived directly down the hill from McQueen and, as McQueen recalled, "I could see that Jim was very neat around his place. Flowers trimmed, no papers in the yard, grass always cut... so, just to piss him off, I'd start lobbing empty beer cans down the hill into his driveway. He'd have his drive all spic 'n span when he left the house, and then get home to find all those empty cans. It took him a long time to figure out it was me."

McQueen was conservative in his political views and often backed the Republican Party. He supported the Vietnam War, was one of the few Hollywood stars to refuse numerous requests to back Presidential hopeful Robert Kennedy in 1968, and turned down the chance to participate in the 1963 March on Washington. When McQueen heard a rumor that he had been added to Nixon's Enemies List, he responded by immediately flying a giant American flag outside his house. His wife Ali McGraw reportedly responded to the whole affair by saying, "But you're the most patriotic person I know."

McQueen was married three times. He married Manila-born actress Neile Adams on November 2, 1956 (divorced 1972), by whom he had a daughter, Terry (born June 5, 1959; died at thirty-eight on March 19, 1998 as a result of hemochromatosis, a condition in which the body produces too much iron destroying the liver), and a son, Chad McQueen (born December 28, 1960 and now an actor, as is his grandson, Steven R. McQueen, born in 1988).

On August 31, 1973 he married his *Getaway* co-star, Ali MacGraw, with whom he had a passionate but tumultuous relationship (she left her husband, film producer Robert Evans, for McQueen). They were divorced in 1978. His third wife was model Barbara Minty, whom he married on January 16, 1980, less than a year before his death.

Steve McQueen died at the age of fifty in Ciudad Juárez, Chihuahua, Mexico from two heart attacks caused by blood clots following a seven-hour operation to remove or reduce a metastatic tumor in his stomach. He had been diagnosed with mesothelioma (a type of cancer associated with asbestos exposure) in December of 1979, and had travelled to Mexico in July of 1980 for unconventional treatment after his doctors advised him they could do nothing more to prolong his life. Controversy arose over McQueen's Mexican trip because he sought a very non-traditional treatment that used coffee enemas and laetrile, a supposedly "natural" anti-cancer drug available in Mexico but not approved by the U.S. Food and Drug Administration.

Steve McQueen was cremated, and his ashes were spread in the Pacific Ocean.

ALVY MOORE
Agent Hank Kimball

Jack Alvin "Alvy" Moore (December 5, 1921 - May 4, 1997) was a light comic actor best known for his role as scatterbrained county agricultural agent Hank Kimball on the television series *Green Acres.*

Born in Vincennes, Indiana, a young Moore moved with his parents to Terre Haute. President of the senior class at Wiley High School, he attended Indiana State Teachers College (now Indiana State University) both before and after service with the Marines during World War II, during which time he saw combat in the Battle of Iwo Jima.

Moore appeared in guest and supporting roles in a number of movies and television shows, including *TheMickey Mouse Club*, where he hosted "What I Want to Be" segments as the "Roving Reporter." He had a small role as a member of Marlon Brando's motorcycle gang in the 1953 film *The Wild One*, which also featured another Marine named Lee Marvin.

His last appearance on television was a brief guest shot on the sitcom *Frasier*.

Alvy Moore died of heart failure on May 4, 1997 at the age of seventy-five.

WARREN OATES
Sergeant Hulka

Warren Mercer Oates (July 5, 1928 - April 3, 1982) was an actor who was born and raised in Depoy, Kentucky, the son of Sarah Alice (née Mercer) and Bayless E. Oates, who owned a general store. He is best known for his performances in several films directed by fellow Marine Sam Peckinpah, most notably *The Wild Bunch* and *Bring Me the Head of Alfredo Garcia*. He also starred in numerous films during the early 1970s which have since achieved cult status including *The Hired Hand, Two-Lane Blacktop* and *Race with the Devil*. Oates also portrayed 'Sergeant Hulka' in the 1981 box office hit *Stripes*.

Oates attended high school in Louisville and enlisted in the Marines in the 1950s. He began his acting career in New York City, starring in a live production of the television series *Studio One* in 1957. The actor migrated to Los Angeles where he began to carve out a niche playing guest roles on the western television programs of the period including *Wagon Train, Tombstone Territory, Rawhide, Wanted: Dead or Alive, Have Gun-Will Travel, The Big Valley* and *Gunsmoke*. Oates first met Peckinpah when he played a variety of guest roles on *The Rifleman*, the popular television series created by the director. He also played a supporting role on Peckinpah's short-lived TV series *The Westerner* in 1960. The collaboration continued as he worked on Peckinpah's early films *Ride the High Country* and *Major Dundee*.

In addition to Peckinpah, Oates worked with several of the major film directors of his era including Norman Jewison in *In the Heat of the Night* , Joseph L. Mankiewicz in *There Was a Crooked Man.*, John Milius in *Dillinger*, Terrence Malick in *Badlands*, Philip Kaufman in *The White Dawn*, William Friedkin in *The Brink's Job*, and Steven Spielberg in *1941*.

His partnership with Peckinpah resulted in two of his most famous film roles. In the 1969 Western classic *The Wild Bunch* he portrayed Lyle Gorch, a longtime outlaw who chooses to die with his friends during the film's violent conclusion. According to his wife at the time, Teddy, Oates had the choice of starring in *Support Your Local Sheriff*, which was to be filmed in Los Angeles, or *The Wild Bunch*, in Mexico. "He had done *Return of the Seven* in Mexico. He got hepatitis, plus the 'revenge.' But off he went again with Sam (Peckinpah). He loved going on location. He loved the adventure of it. He had great admiration for Sam. Sam Peckinpah and Monte Hellman were the two directors Warren would work with anytime, anywhere." In *Bring Me the Head of Alfredo Garcia,* the dark 1974 action/tragedy also filmed in Mexico, Oates played the lead role of Bennie, a hard-drinking down-on-his-luck musician hoping to make a final score. The character was reportedly based on Peckinpah himself, and for authenticity Oates wore the director's sunglasses while filming his scenes.

Although the Peckinpah film roles are his best-known, his most critically acclaimed role is 'GTO' in Monte Hellman's 1971 cult classic *Two-Lane Blacktop*. The film, although a failure at the box-office, is studied in film schools as a treasure of the '70s in large part due to Oates' heartbreaking portrayal of GTO.

A year before his death Oates co-starred with Bill Murray in the 1981 military comedy *Stripes*. In the role of rigid drill sergeant 'Sergeant Hulka,' Oates skillfully played the straight man to Murray's comedic character. The film was a huge financial success, earning eighty-five million at the box office.

Warren Oates died of a sudden heart attack in Los Angeles, California on April 3, 1982 at the age of fifty-three, and his last two films, *Blue Thunder* and *Tough Enough* (both released in 1983), were posthumously dedicated to him. Oates was cremated, and his ashes were scattered in Montana.

Today he has a dedicated cult following due to his memorable performances in not only Peckinpah's films, but also Monte Hellman's independent works, his films with Peter Fonda, and a number of B-movies from the 1970s. His occasionally crude facade, likeable persona, and uncommon presence are admired by such filmmakers as Quentin Tarantino and Richard Linklater. During a recent screening of Hellman's *Two-Lane Blacktop* Linklater introduced the film and announced sixteen reasons viewers should love the 1971 movie. His sixth reason was, "Because there was once a god who walked the earth named Warren Oates."

HUGH O'BRIAN
Wyatt Earp

"I believe every person is created as the steward of his or her own destiny with great power for a specific purpose."
- Hugh O'Brian

Hugh O'Brian, born Hugh Charles Krampe, is an actor who was born on April 19, 1925 in Rochester, New York. He is best known for his starring role as Wyatt Earp in the ABC television series *The Life and Legend of Wyatt Earp* which aired from 1955 until 1961.

O'Brian attended New Trier High School in Winnetka, Illinois (as did Rock Hudson, Charlton Heston, Ann-Margret and many other future stars) and later Kemper Military School in Boonville, Missouri. In high school he lettered in football, basketball, wrestling and track. After a semester at the University of Cincinnati with studies charted toward a law career O'Brian, at the age of seventeen, enlisted in the Marine Corps in 1942.

Following World War II O'Brian moved to Los Angeles and found work on stage and in film. He got his big break when he was chosen to portray the legendary lawman Wyatt Earp on television. *The Life and Legend of Wyatt Earp* debuted in 1955 as the "first adult western" and it soon became one of the top-rated shows on television. During its seven-year run *Wyatt Earp* consistently placed in the top ten in the United States. He also appeared regularly on other programs in the 1960s. For example, he was a guest panelist on the popular Sunday night CBS program *What's My Line?* and served as a mystery guest three times.

The actor made a number of motion pictures, among them *The Lawless Breed, There's No Business Like Show Business* and *In Harm's Way.* He was also a featured star in the two-hour premiere of television's *Fantasy Island.* Perhaps O'Brian's greatest distinction is he is the last man John Wayne ever killed on screen in his final movie *The Shootist* in 1976. O'Brian was a good friend of the Duke, and said he considered it a great honor.

O'Brian recreated his Wyatt Earp role for two 1990s projects, *Guns of Paradise* and *The Gambler Returns: Luck of the Draw* with fellow actor Gene Barry doing likewise as lawman Bat Masterson for each. He also had a small role in the Danny DeVito/Arnold Schwarzenegger comedy *Twins* as one of several men who had "donated" the DNA that later became the "twins." In the film Schwarzenegger thought he'd found his "father" when he met Hugh O'Brian's character.

On June 25, 2006 O'Brian, at the age of eighty-one, married for the first time. His wife is the former Virginia Barber (she was born in 1952). The ceremony was held at Forest Lawn Memorial Park with the Reverend Robert Schuller, pastor of the Crystal Cathedral in Garden Grove, officiating. The couple was serenaded by close friend Debbie Reynolds.

Hugh O'Brian has dedicated much of his life to the Hugh O'Brian Youth Leadership (HOBY). HOBY is a non-profit youth leadership development program that empowers ten thousand sophomores annually through its over seventy leadership programs in all fifty states and eight countries. Since its inception in 1958 over 355,000 young people have been inspired by HOBY.

One high school sophomore from every high school in the United States, referred to as an "ambassador," is welcome to attend a state or regional HOBY seminar. From each of those seminars, two students are offered the opportunity to attend the World Leadership Conference (WLC). Many do not attend because it is quite expensive, but several funds and scholarships, such as the Jack Tawney Memorial Fund in the Central PA chapter, allow students to go for free.

The concept for HOBY was inspired in 1958 by a nine-day visit O'Brian had with famed humanitarian Dr. Albert Schweitzer in Africa. Dr. Schweitzer believed "the most important thing in education is to teach young people to think for themselves."

Hugh O'Brian's message to young people is "Freedom to Choose." He has said, "I do NOT believe we are all born equal. Created equal in the eyes of God yes, but physical and emotional differences, parental guidelines, varying environments, being in the right place at the right time, all play a role in enhancing or limiting an individual's development. But I DO believe every man and woman, if given the opportunity and encouragement to recognize their potential, regardless of background, has the freedom to choose in our world. Will an individual be a taker or a giver in life? Will that person be satisfied merely to exist, or seek a meaningful purpose? Will he or she dare to dream the impossible dream? I believe every person is created as the steward of his or her own destiny with great power for a specific purpose, to share with others, through service, a reverence for life in a spirit of love."

For his contribution to the television industry Hugh O'Brian has a star on the Hollywood Walk of Fame at 6613-1/2 Hollywood Blvd, and in 1992 he was inducted into the Western Performers Hall of Fame at the National Cowboy & Western Heritage Museum in Oklahoma City, Oklahoma.

GERALD S. O'LOUGHLIN
The Rookies

"My Marine Corps experience means a great deal to me in many ways. The camaraderie, discipline and tenacity I learned helped me face some awful situations in my life."
–Gerald S. O'Loughlin

Gerald Stuart O'Loughlin, Jr. is a television, stage and film actor who was born in New York City on December 23, 1921. Short and dark, but tough-talking and rough-looking, he received his start on the stage after pondering a career in law. After a stint with the Marine Corps he used his GI bill to train in New York at the Neighborhood Playhouse. Throughout the early 50s he was frequently seen in TV dramas and highlighted his stage career with a national tour of *A Streetcar Named Desire* as Stanley Kowalski with the incomparable Tallulah Bankhead starring as Blanche DuBois, and the role of mental patient Cheswick opposite Kirk Douglas' Randle McMurphy in 1963's *One Flew Over the Cuckoo's Nest* on Broadway. He made little impression in films at the beginning with an offbeat romantic lead role in the low budgeted *Lovers and Lollipops* and a small role in the more impressive *A Hatful of Rain*, but later toughened things up a bit with parts in *In Cold Blood, Ice Station Zebra*, and especially *Desperate Characters*. Things finally came together for him on 70's TV when he nabbed the role of Lt. Ryker in the police series *The Rookies*, replacing Darren McGavin, who had played the same role in the pilot. O'Loughlin was successful in mini-movies as well, especially as the patriarch in the tearjerker *Something for Joey* with Geraldine Page. He played in other less successful TV series such as *Automan*, and continued acting into the millennium, albeit less and less.

He is divorced from casting director Meryl Abeles, who went by the name Meryl O'Loughlin. They have two children, Chris and Laura. Chris was a member of the 1992 U.S. Olympic team in épée fencing.

PAT PAULSEN
Savior of America's Destiny!

Patrick Layton Paulsen (July 6, 1927- April 24, 1997) was a comedian and satirist who was born in South Bend, Washington. He is most notable for his roles on several of the *Smothers Brothers* TV shows, and for his campaigns for President of the United States in 1968, 1972, 1980, 1988, 1992, and 1996 which had primarily comedic rather than political objectives - although his campaigns generated some protest votes for him.

After graduating from Tamalpais High School in Mill Valley, Paulsen immediately joined the Marines during World War II. He returned home after the war and worked several jobs, including working as a postal clerk, truck driver, selling Fuller brushes, and toiling in a gypsum mine. Later he was employed as a photostat operator for several years. After attending San Francisco City College Paulsen joined an acting group called "The Ric-y-tic Players" and formed a comedy trio which included his brother Lorin.

157

Paulsen went on to become a single act, appearing as a comedic guitarist in various clubs on the west coast and in New York City. During one of his appearances in San Francisco he met the Smothers Brothers.

In 1967 *The Smothers Brothers Comedy Hour* premiered. Paulsen said he was hired because he sold them cheap songs and would run errands. At first he was cast as their editorialist, and his deadpan, double-talk comments on the issues of the day propelled him into the national consciousness. His deadpan work was nearly flawless. On one isolated occasion, in a talk about Hawaii, he defined a 'wahine' as something you put on a 'bu-hun' with lots of 'mu-hustard.' His composure started to crack, but he recovered. His work on *The Smothers Brothers Comedy Hour* earned Paulsen an Emmy in 1968. Early in 1970 Paulsen headlined his own series, *Pat Paulsen's Half a Comedy Hour*, which ran thirteen weeks on ABC. Guests on the first show were former Vice President of the United States Hubert Humphrey and an animated Daffy Duck, the latter of whom was interviewed by Paulsen.

The comedian was approached by the Smothers Brothers with the idea of running for President in 1968. His reply, he was later to recount, was "Why not? I can't dance. Besides, the job has a good pension plan and I'll get a lot of money when I retire."

Paulsen's campaign that year, and in succeeding years, was grounded in comedy – although it was not bereft of serious commentary. He ran the supposed campaigns using obvious lies, double talk and tongue-in-cheek attacks on the major candidates, and responded to all criticism with his catch phrase "Picky, picky, picky." His campaign slogan was "Just a common, ordinary, simple savior of America's destiny." A good example of his thinking was the quip,

"Why should we tell kidnappers, murderers and embezzlers their rights? If they don't *know* their rights, they shouldn't be in the business!"

Paulsen's name actually appeared on the ballot in New Hampshire for the Democratic Primary several times, and in 1996 he received 921 votes (1%) to finish second to President Bill Clinton (76,754 votes). In 1992 he came in second to George Bush in the North Dakota Republican Primary, and in the 1992 Republican Party presidential primary he received 10,984 votes in total.

During later years Paulsen appeared in nightclubs, theaters, and conventions throughout the country. He also appeared each summer in Traverse City, Michigan at the Cherry County Playhouse, where he produced and starred in some twenty-five different plays including *The Fantasticks, The Odd Couple, Harvey*, and *The Sunshine Boys*.

Pat Paulsen died of complications from colon and brain cancer and pneumonia in Tijuana, Mexico on April 25, 1997 at the age of sixty-nine.

SAM PECKINPAH
Bloody Sam

"The Corps is not a finishing school. It is not a game. It's a way of life – and its basic purpose is to maintain the fact that a combat Marine is the best fighting soldier in the world." – *Sam Peckinpah, in a letter to his son after the latter joined the Marine Corps*

David Samuel "Sam" Peckinpah (February 21, 1925 - December 28, 1984) was a film director who was born in Fresno, California where he attended both grammar school and high school but spent much of his time skipping classes with his brother to engage in cowboy activities such as trapping, branding and shooting. He achieved iconic status following the release of his 1969 Western epic *The Wild Bunch,* and became one of the major filmmakers of the 1970s with his innovative and explicit depiction of action and violence, as well as his revisionist approach to the Western genre.

Peckinpah's films generally deal with the conflict between values and ideals and the corruption of violence in human society. He was given the nickname "Bloody Sam" due to the violence in his films. His characters are often loners or losers who desire to be honorable, but are forced to compromise in order to survive in a world of nihilism and brutality.

He played on the junior varsity football team while at Fresno High School but frequent fighting and discipline problems caused his parents to enroll him in the San Rafael Military Academy for his senior year, and in 1943 he joined the Marine Corps. Within two years his battalion was sent to China with the task of disarming Japanese soldiers and repatriating them following World War II, and while his duty did not include combat he claimed to have witnessed acts of war between Chinese and Japanese soldiers. According to friends, these included several acts of torture and the murder of a laborer by random sniper fire - but the American Marines were not permitted to intervene. This reportedly deeply affected Peckinpah, and may have influenced his depictions of violence in his films.

After the war he attended Fresno State College where he studied history. While a student he met and married his first wife, Marie Selland, in 1947. A drama major, Selland introduced Peckinpah to the theatre department and he became interested in directing for the first time. During his senior year he adapted and directed a one-hour version of Tennessee Williams' *The Glass Menagerie*. After graduation in 1948, Peckinpah enrolled in graduate studies in drama at the University of Southern California. He spent two seasons as the director in residence at Huntington Park Civic Theatre near Los Angeles before obtaining his Master's degree. He was asked to stay on another year, but Peckinpah began working as a stagehand at KLAC-TV in the belief that television experience would eventually lead to work in films.

Even during this early stage of his career Peckinpah was developing a combative streak. He was reportedly kicked off the set of *The Liberace Show* for not wearing a tie, and refused to cue a car salesman during a live feed because of his attitude towards stagehands.

In 1954 Peckinpah was hired as "dialogue director" for the film *Riot in Cell Block 11* and his job entailed acting as a 'gopher' for the movie's director, Don Siegel. The movie was filmed on location at Folsom Prison and the warden was reluctant to allow the filmmakers to work there until he was introduced to Peckinpah. The warden knew his family from Fresno, and immediately became cooperative. Siegel's location work and his use of actual prisoners as extras in the film made a lasting impression on Peckinpah. He would work as an assistant to the director on four additional films including *Private Hell 36, An Annapolis Story, Invasion of the Body Snatchers* and *Crime in the Streets. Body Snatchers,* in which Peckinpah appeared in a cameo as Charlie the meter reader, starred Kevin McCarthy and Dana Wynter. It would become one of the most critically praised science fiction films of the 1950s.

An experienced hunter, Peckinpah was fascinated with firearms and was known to shoot the mirrors in his house while abusing alcohol - and this image occurred several times in his films. Peckinpah's reputation as a hard-living brute with a taste for violence, inspired by the content in his most popular films and in many ways perpetuated by himself, has overshadowed his artistic legacy. His friends and family have claimed this does a disservice to a man who was actually more complex than generally credited. Throughout his career Peckinpah inspired extraordinary loyalty in certain friends and employees. He used the same actors (L. Q. Jones, R. G. Armstrong, Kris Kristofferson,

James Coburn, Ben Johnson, and fellow Marine Warren Oates) and collaborators in many of his films, and several of his friends and assistants stuck by him to the end of his life.

After divorcing Selland, the mother of his first four children, in 1960, he would eventually marry Mexican actress Begoña Palacios in 1965. A stormy relationship developed, and over the years they would marry three separate times. Peckinpah spent a great deal of his life in Mexico after his marriage to Palacios, eventually buying property there. He was reportedly fascinated by the Mexican lifestyle and culture, and often portrayed it with an unusual sentimentality and romanticism in his films. Four of his films, *Major Dundee, The Wild Bunch, Pat Garrett & Billy the Kid* and *Bring Me the Head of Alfredo Garcia*, were filmed entirely on location within the country while *The Getaway* concludes with a couple escaping to freedom in Mexico.

In 1958 Peckinpah wrote a script for *Gunsmoke* that was rejected due to content. He reworked the screenplay, titled *The Sharpshooter*, and sold it to Zane Grey Theater. The episode received popular response and became the television series *The Rifleman* starring Chuck Connors. Peckinpah would direct four episodes of the series (with guest stars R. G. Armstrong and Warren Oates), but left after the first year. *The Rifleman* would run for five seasons and achieve enduring popularity in syndication.

During this time he also created the television series *The Westerner,* starring John Dehner and fellow Marine Brian Keith. From 1959 to 1960 Peckinpah acted as producer of the series, having a hand in the writing of each episode and directing five of them. Critically praised, the show ran for only thirteen episodes before cancellation mainly due to its gritty content detailing the drifting, laconic cowboy Dave

Blassingame (Keith). Despite its short run, *The Westerner* and Peckinpah would be nominated by the Producers Guild of America for Best Filmed Series. An episode of the series eventually served as the basis for Tom Gries' 1968 film *Will Penny*. *The Westerner*, which has since achieved cult status, further established Peckinpah as a talent to be reckoned with.

In 1967 Warner Brothers' producers Kenneth Hyman and Phil Feldman were interested in having Peckinpah rewrite and direct an adventure film. An alternative screenplay written by Roy Sickner and Walon Green was the western *The Wild Bunch*. At the time, William Goldman's screenplay *Butch Cassidy and the Sundance Kid* – which would be directed by fellow Marine George Roy Hill - had recently been purchased by 20th Century Fox.

It was quickly decided that *The Wild Bunch*, which had several similarities to Goldman's work, would be produced in order to beat *Butch Cassidy* to the theaters. By the fall of 1967 Peckinpah was rewriting the screenplay into what would become *The Wild Bunch*. Filmed on location in Mexico, Peckinpah's epic work was inspired by many things - his hunger to return to films, the violence seen in Arthur Penn's Bonnie and Clyde, America's growing frustration with the Vietnam War, and what he perceived to be the utter lack of reality seen in Westerns up to that time. He set out to make a film which portrayed not only the vicious violence of the period, but the crude men attempting to survive the era. Starring William Holden, Ernest Borgnine, Ben Johnson, Warren Oates, Edmond O'Brien, and Robert Ryan – yet another former Marine - the film detailed a gang of veteran outlaws on the Texas/Mexico border in 1913 trying to exist within a rapidly approaching modern world. *The Wild Bunch* is framed by two ferocious and infamous gunfights, beginning with a failed bank robbery and concluding with the outlaws battling the Mexican army in suicidal vengeance due to the death of one of their members. Irreverent and unprecedented in its explicit detail, the 1969 film was an instant classic. Many critics denounced its violence as sadistic and exploitative. Other critics and filmmakers hailed the originality of its unique rapid editing style, created for the first time in this film and ultimately becoming a Peckinpah trademark, and praised the reworking of traditional Western themes. It was the beginning of Peckinpah's international fame, and he and his work would remain controversial for the rest of his life. The film would be ranked number eighty on the American Film Institute's top one hundred list of the greatest American films ever made, and number sixty-nine as the most thrilling - but the controversy has not diminished.

In May of 1971 he returned to the United States to begin work on *Junior Bonner*. Filmed on location in Prescott, Arizona and starring another Marine Corps alumnus (Steve McQueen), the story covered a week in the life of aging rodeo rider Junior "JR" Bonner as he returns to his hometown to compete in an annual rodeo competition. In addition to McQueen the cast included Robert Preston, Joe Don Baker, Ben Johnson and Ida Lupino – the ex-wife of former Marine Louis Hayward. *Junior Bonner* was marked by sharp character development, colorful location detail and unusually tender scenes between Preston and Lupino as Bonner's estranged parents. Promoted as a Steve McQueen action vehicle, reviews were mixed and the film performed poorly at the box office. Peckinpah would remark, "I made a film where nobody got shot, and nobody went to see it." The film's reputation has grown over the years, and many critics now consider *Junior Bonner* to be one of Peckinpah's most sympathetic works.

Stinging from the failure of *Junior Bonner* but eager to work with Peckinpah again, Steve McQueen presented him Walter Hill's screenplay to *The Getaway*. Based on the Jim Thompson novel, the gritty crime thriller detailed lovers on the run following a dangerous robbery. Both Peckinpah and McQueen needed a hit, and they immediately began working on the film in February of 1972. Peckinpah had no pretensions about making *The Getaway*, as his only goal was to create a highly-polished thriller to boost his market value. McQueen would play Doc McCoy, an imprisoned mastermind robber whose wife Carol (Ali MacGraw) conspires for his release on the condition they rob a bank in Texas. A double cross follows the crime, and the McCoy's are forced to flee for Mexico with both the police and criminals in hot pursuit. Replete with explosions, car chases and intense shootouts, the film would become Peckinpah's biggest financial success to date, earning more than twenty-five million at the box office.

Conflicts of masculinity were a major theme of his work, leading some critics to compare him to Ernest Hemingway. Peckinpah's world is a *man's* world, and feminists have castigated his films as misogynistic and sexist, especially the shooting of a woman during the final moments of *The Wild Bunch,* the rape sequence in *Straw Dogs* and Doc McCoy's physical assault of his wife in *The Getaway*. His defenders point out that while the women in his films are generally seen through men's eyes, it is the men who are abusive, corrupted and violent. The women are either victims of the brutalities of men, or survivors attempting to eke out an existence in the unforgiving world created by men. Peckinpah saw violence as the product of human society, and not of nature. It is the result of men's competition with each other over power and domination, and their inability to negotiate this competition without resorting to brutality.

Peckinpah, who was born to a ranching family that included judges and lawyers, was also deeply concerned by the conflict between "old-fashioned" values and the corruption and materialism of the modern world. Many of his characters attempt to live up to their expectations of themselves, even as the world they live in demands that they compromise their values. This is most explicitly stated in the famous exchange from *Ride the High Country* in which Steve Judd (Joel McCrea) states, "All I want is to enter my house justified." Many believe that this line is taken directly from a common expression used by David Peckinpah, the director's father.

In many ways Peckinpah's greatest legacy lies in his aggressive breaking of taboos. He allowed a new freedom to emerge in cinema, not only in the depiction of violence, but also in editing styles, narrative choices, and the willingness to portray unsympathetic or tragic characters and stories – but unfortunately his notorious reputation has often overshadowed the depth of his influence on modern film.

Peckinpah was seriously ill during the final years of his life as a lifetime of self-abuse began to catch up with him. Regardless, he continued to work until the last months before his death, until he died of heart failure on December 28, 1984while in preparation for shooting an original script by Stephen King entitled *The Shotgunners.*

GEORGE PEPPARD
Hannibal Smith

George Peppard, Jr. (October 1, 1928 - May 8, 1994) was a well known film and television actor who was born to building contractor George Peppard, Sr. and opera singer Vernelle Rohrer in Detroit, Michigan.

Peppard secured a major role early in his career when he starred alongside Audrey Hepburn in the 1961 movie *Breakfast at Tiffany's,* but he is probably best known to younger audiences for his role as Colonel John "Hannibal" Smith on the 1980s television show *The A-Team,* where he was the cigar-chomping leader of a renegade commando squad, and as the millionaire sleuth on the show *Banacek.*

Peppard enlisted in the Marine Corps at age seventeen and rose to the rank of acting Gunnery Sergeant in the artillery, but left the Marines at the end of his first tour. He then studied Civil Engineering at Purdue University, where he was a member of Beta Theta Pi. He also attended Carnegie Mellon University in Pittsburgh, Pennsylvania.

Peppard made his stage debut in 1949 at the Pittsburgh Playhouse. He then enrolled in The Actor's Studio in New York. His first work on Broadway led to his first television appearance in 1956 along with a young Paul Newman in *The United States Steel Hour.* Peppard's Broadway appearance in *The Pleasure of His Company* followed, and led to an MGM contract. Prior to that he had a strong film debut in *The Strange One*, and he was discovered playing with Robert Mitchum's illegitimate son in the popular melodrama *Home From the Hill*.

His good looks, elegant manner and superior acting skills landed Peppard his most famous film role as Paul Varjak (aka "Fred") in *Breakfast at Tiffany's* with Audrey Hepburn. This role boosted him briefly to being a major film star. His leading roles included *How the West Was Won* in 1962, *The Carpetbaggers* in 1964 and *The Blue M*ax in 1966.

Peppard developed a tendency to choose tough guy roles in big, ambitious pictures where he was somewhat overshadowed by ensemble casts. A good example is his role as German pilot Bruno Stachel, an obsessively competitive officer from humble beginnings who challenges the Prussian aristocracy during World War I, in *The Blue Max*. For that role Peppard learned to fly, earned a private pilot's license, and did his own stunt flying.

Due to Peppard's tendencies toward alcohol his career led to a string of B films, except for a brief moment of notable success with the highly successful TV series *Banacek*, which ran from 1972 to 1974 as part of the *NBC Mystery Movie* series, and one of his most critically acclaimed, though rarely seen, performances in the TV movie *Guilty or Innocent: The Sam Sheppard Murder Case*.

Among his most disappointing films was the 1970 western *Cannon for Cordoba*, in which Peppard played the steely Captain Rod Douglas, who had been put in charge of gathering a group of soldiers for a dangerous mission into Mexico, and 1967's *Rough Night in Jericho* in which he co-starred with Dean Martin. Peppard also appeared in the short lived (only 1/2 season) *Doctors' Hospital* in 1975, and the science fiction film *Damnation Alley* in 1977. Then, with few interesting film roles coming his way, he acted in, directed and produced the drama *Five Days from Home* in 1979.

In a rare game show appearance Peppard did a week of shows on *Password Plus* in 1979. Five shows were filmed, but one was never broadcast on NBC (but aired *much* later on GSN) due to a rant where Peppard expressed dissatisfaction with NBC executives watching "as if you're some sort of crook." He was never asked to return.

Peppard was notably offered the original role of Blake Carrington in the TV series *Dynasty*, and actually filmed the pilot episode with Linda Evans and Bo Hopkins. He later turned down the role due to disagreements with writers, and the part was subsequently offered to John Forsythe with the scenes with Peppard being reshot.

In the early 1980s George Peppard re-emerged as a television star for his role as "Hannibal" Smith in the action adventure series *The A-Team*, acting alongside Mr. T, Dirk Benedict and Dwight Schultz. In the series the A-Team was a crack squad of renegade commandos on the run from the military for a crime they did not commit while serving in the Vietnam War. The Team made a living as soldiers of fortune, albeit only helping people with a just grievance.

"Hannibal" Smith was the leader of the A-Team, distinguished by his cigar-smoking, black leather gloves, disguises, and catch phrase, "I love it when a plan comes together!" The show ran five seasons on NBC from 1983 until 1987, made Peppard known to a younger generation, and is arguably his most well-known role. The part was reportedly written with James Coburn in mind, but it went to Peppard when Coburn had to pull out.

The actor was married five times to Helen Davies, Elizabeth Ashley, Sherry Boucher-Lytle, Alexis Adams and Laura Taylor. He was also the father of three children.

Peppard gave up drinking in 1978 and spent his later years trying to assist other alcoholics with their recovery. A life-long smoker, he was diagnosed with lung cancer in 1992 and subsequently quit that as well. His fifth wife and "number one fan," former West Palm Beach banker Laura Taylor, met and married him shortly after he was diagnosed with terminal cancer and nursed him through his last eighteen months. Cancer never forced his retirement from acting however, and Peppard completed a pilot for a new series in 1994 (a *Matlock* spin-off) shortly before his passing.

George Peppard died on May 8, 1994 in Los Angeles, California. Although he was being treated for cancer his actual cause of death was a complication arising from the treatment he was getting - chemotherapy-induced leukemia. He is buried in Northview Cemetery in Dearborn, Michigan.

TYRONE POWER
"Good Night, Sweet Prince"

Tyrone Edmund Power, Jr. (May 5, 1914 - November 15, 1958) was a film and stage actor who appeared in dozens of films from the 1930s to the 1950s, often in swashbuckler roles or romantic leads such as *The Mark of Zorro, The Black Swan, Prince of Foxes, The Black Rose,* and *Captain from Castile.* Though famous for the dark, classically handsome looks which made him a matinee idol from his first film appearance, Power was also very versatile. He played a wide range of roles from a protagonist with a darker side to light romantic comedy, and in the 1950s began placing limits on the number of movies he would make in order to have time for the stage. There he received his biggest accolades in *John Brown's Body* and *Mister Roberts.*

Born in Cincinnati, Ohio in 1914, the only son of English-born American stage and screen actor Tyrone Power, Sr. and Helen Emma "Patia" Reaume, Power was descended from a long theatrical line going back to his great-grandfather, Irish born actor and comedian Tyrone Power (1795-1841). During the first year of Power's life he lived in Cincinnati, and his father was absent for long periods due to his stage commitments in New York. Young Power was a sickly child, and his doctor advised his family that the climate in California might be better for his health. The family moved there in 1915, and his parents appeared together on stage and in 1917 in the movie *The Planter.* Thereafter Tyrone Power, Sr., as he later became known, found himself away from home more frequently, as his stage career often took him to New York. The Powers eventually drifted apart and divorced around 1920.

After the divorce Patia Power worked as a stage actress and in 1921 (at the age of seven) young Tyrone appeared with his mother in the mission play *La Golondrina.* A couple of years later the family moved back to Cincinnati, where Power graduated from Purcell High School in 1931. Then, upon his graduation, he opted to join his father to learn what he could about acting from one of the stage's most respected actors.

Power joined his father for the summer of 1931 after being separated from him for some years due to his parents' divorce, but his father suffered a heart attack in December of 1931 and died in his son's arms while preparing to perform in *The Miracle Man.*

Sometime later Power was in Chicago where his friend, radio personality Don Ameche, convinced him to stay a while to work in radio. He wasn't able to get a foothold however, and eventually went on to New York. There he met the great stage actress Katharine Cornell, who cast him as an understudy for Burgess Meredith in the play *Flowers of the Forest.* A better stage break came when Cornell put him in the role of Benvolio in *Romeo and Juliet*. Hollywood scouts saw him and offered a screen test. Katharine Cornell advised against going to Hollywood without a little more stage experience, and Power took her advice. Cornell then gave him a substantial role in her next stage play, *St. Joan*. Once again Hollywood scouts saw him and offered a screen test, and this time Cornell told him he was ready.

Tyrone Power went to Hollywood in 1936, where he signed with 20th Century-Fox. He would be their top leading man for years to come. Director Henry King was impressed with his looks and poise and insisted Power be tested for the lead role in *Lloyd's of London*, a role thought to belong to Don Ameche. Despite Darryl F. Zanuck's reservations he decided to go ahead and give him the lead role in the movie, once Henry King and Fox editor Barbara McLean convinced him Power had a greater screen presence than Ameche. He was fourth billed in the movie, but had by far the most screen time. Tyrone Power walked into the premiere of the movie an unknown, and walked out a superstar - where he stayed for the remainder of his career.

Power racked up hit after hit from 1936 until 1943, when his career was interrupted by military service. In 1940 the direction of his career took a dramatic turn when his movie, *The Mark of Zorro*, was released. Power played the role of Don Diego Vega, fop by day and bandit hero Zorro by night. The role had been made famous by Douglas Fairbanks in the 1920 movie by the same title. Power's performance was excellent, and 20th Century Fox often cast him in swashbucklers in the years that followed. Power was actually an excellent swordsman, and the dueling scene in *The Mark of Zorro* is considered one of the finest in screen history. The great Hollywood swordsman Basil Rathbone, who starred with him in *The Mark of Zorro*, commented, "Power was the most agile man with a sword I've ever faced before a camera. Tyrone could have fenced Errol Flynn into a cocked hat."

Despite being kept busy making movies at 20th Century-Fox, Power also found time to do radio and stage work. He appeared with his wife Annabell, in several radio broadcasts, including the plays *Blood and Sand, The Rage of Manhattan,* and *Seventh Heaven*. He also worked with other big names in radio, including Humphrey Bogart, Jeanne Crain, Loretta Young, Alice Faye, and Al Jolson.

In August 1942 Power enlisted in the Marine Corps. He reported to the Marines for training in late 1942, but was sent back at the request of 20th Century-Fox to complete one more film, 1943's Crash Dive - a patriotic war movie. He was credited in the movie as Tyrone Power, USMCR, and the movie served as much as anything as a recruiting film. Other than re-releases of his films, he wasn't seen on screen again until 1946 when he co-starred with Gene Tierney in *The Razor's Edge*, an adaptation of Somerset Maugham's novel of the same name.

After completing boot camp at Marine Corps Recruit Depot San Diego he attended Officer's Candidate School at Marine Corps Base Quantico, where he was commissioned a Second Lieutenant on June 2, 1943. Because he had already logged 180 solo hours as a pilot prior to enlisting in the Marine Corps Power was able to go through a short, intense flight training program at Naval Air Station Corpus Christi, Texas where he earned his wings and was promoted to First Lieutenant. He arrived at Marine Corps Air Station Cherry Point, North Carolina in July of 1944 and was assigned to VMR-352 as an R5C copilot. When the squadron moved to Marine Corps Air Station El Toro in California in October of 1944 Power was reassigned to VMR-353, and joined them on Kwajalein in February of 1945 where he flew cargo and wounded Marines during the Battles of Iwo Jima and Okinawa. He returned to the United States in November of 1945, and he was released from active duty in January 1946. Power was promoted to Captain in the reserves on May 8, 1951 but was not recalled for service during the Korean War.

In the June 2001 *Marine Air Transporter* newsletter Jerry Taylor, a retired Marine Corps flight instructor, recalled memories of World War II. He spoke of training Tyrone Power as a pilot, saying, "He was an excellent student, never forgot a procedure I showed him, or anything I told him." Others have also commented that he was well-respected by those with whom he served.

As the 1950s rolled around Power became increasingly unhappy with his movie assignments such as *American Guerrilla in the Philippines* and *Pony Soldier* and asked his studio for permission to seek out his own roles. Permission was granted with the understanding that he would fulfill his fourteen-film commitment to 20th Century-Fox in between those roles. In 1953 he made *The Mississippi Gambler* for Universal Studios, worked a deal to get a percentage of the profits, and ended up making one-million dollars from the movie - a very large sum in those days.

Tyrone Power's last movie made under his contract with 20th Century-Fox was released in 1955. That same year *The Long Gray Line*, a hugely successful John Ford film, was released by Columbia Pictures. Columbia also released *The Eddy Duchin Story*, also huge at the box office, the following year. His old boss, Darryl F. Zanuck, then pressed him into service for the lead role in 1957's *The Sun Also Rises*, adapted from the Ernest Hemingway novel. Tyrone Power's last role turned out to be one of his most highly regarded. He was cast against type as the accused murderer, Leonard Vole, in Agatha Christie's *Witness for the Prosecution.* Critic Robert Fulford of *The National Post* commented on the "superb performance" of Power as "the seedy, stop-at-nothing exploiter of women," which was in sharp contrast to his earlier swashbuckling roles and romantic heroes. The movie was critically acclaimed and was a box office success.

In September of 1958 Tyrone Power went to Madrid and Valdespartera, Spain to film the epic *Solomon and Sheba*, and had filmed about seventy-five percent of his scenes when he was stricken with a massive heart attack as he was filming a dueling scene with his frequent co-star and friend, George Sanders. He died enroute to the hospital, and Yul Brynner was brought in to take over the role of Solomon. The filmmakers still used some of the long shots that Power had filmed, and an observant fan can see him in some of the scenes, particularly in the middle of the duel.

Tyrone Power's last movie was to be in a familiar role - with sword in hand. He is perhaps best remembered as a swashbuckler, and was indeed one of the finest swordsmen in Hollywood. Director Henry King said, "People always seem to remember Ty with sword in hand, although he once told me he wanted to be a character actor. He actually was quite good - among the best swordsmen in films."

Tyrone Power was honored by having his handprints and footprints put in cement at Grauman's Chinese Theater on May 31, 1937 in a joint ceremony with Loretta Young on the occasion of the premiere of their movie *Cafe Metropole*. At the time of the ceremony Power was just twenty-three years old and had been a major star for only six months. He signed the cement block, "To Sid - Following in my father's footsteps," which was a tribute to his father.

When Tyrone Power died at the age of forty-four he was buried at Hollywood Cemetery, now known as Hollywood

Forever Cemetery, at noon in a military service. The active pallbearers were officers of the Marine Corps, and the honorary pallbearers were Charles Laughton, Raymond Massey, Tommy Noonan, Theodore Richmond, Murray Steckler, Cesar Romero, Watson Webb, Milton Bren, James Denton, George Sidney, George Cohen, Lew Schreiber, Lew Wasserman, and Harry Brand. Cesar Romero gave the eulogy, using in it a tribute written by Tyrone Power's good friend and frequent co-star, George Sanders. Sanders had written the tribute on the set of *Solomon and Sheba* within hours of Power's death. It read as follows: "I shall always remember Tyrone as a bountiful man, a man who gave freely of himself. It mattered not to whom he gave. His concern was in the giving. I shall always remember his wonderful smile, a smile that would light up the darkest hour of the day, like a sunburst. I shall always remember Tyrone Power as a man who gave more of himself than it was wise for him to give until, in the end, he gave his life."

Flying over the service was Henry King, who had directed him in eleven movies. Almost twenty years earlier Power had flown with King, in King's plane, to the set of *Jesse James* in Missouri. It was then that Tyrone Power got his first experience with flying, which would become such a big part of his life, both in the Marines and in private life. In the foreword to Dennis Belafonte's book *The Films of Tyrone Power* King said, "Knowing his love for flying and feeling that I had started it, I flew over his funeral procession and memorial park during his burial and felt that he was with me."

Tyrone Power was laid to rest by a small lake in one of the most beautiful parts of the cemetery. His grave is marked by a unique tombstone, in the form of a marble bench. On the tombstone are the masks of comedy and tragedy, with the inscription, "Good night, sweet prince."

As a footnote, Tyrone Power's will contained an unusual provision. It stated his wish that, upon his death, his eyes were to be donated to the Estelle Doheny Eye Foundation for such purposes as the trustees of the foundation should deem advisable - including transplantation of the cornea to the eyes of a living person.

HARI RHODES
A Chosen Few

Hari Rhodes (April 10, 1932 - January 15, 1992) was an author and actor whose career spanned three decades beginning around 1960. He was sometimes billed as Harry Rhodes, and appeared in sixty-six films or television programs such as ABC's medical drama about psychiatry, *Breaking Point.*

In a 1968 *TV Guide* interview Rhodes described growing up in a rough section of Cincinnati. "We lived between the railroad tracks and the river bank. The flood ran us out every winter but we'd always come back, kick out the mud and settle down again until flood time. All the boys had to learn how to hop freights and throw pieces of coal off. All I ever knew was rats, roaches and poverty."

When he was fifteen Rhodes spent two months learning to copy his mother's signature and forged it on enlistment papers in order to join the Marine Corps.

In the Marines Rhodes was a member of his camp's judo team for two years, eventually gained the rank of sergeant and served in Korea where he led a reconnaissance platoon behind enemy lines.

"The time I got wounded at the Chosin Reservoir, a Chinese came running toward me," Rhodes told *TV Guide*. "My Thompson submachine gun was unloaded. I threw it down so he wouldn't shoot. His face almost smiled. He had his bayonet on my chest. He began slashing my arms. I got him with an 8-inch knife."

His most notable role was on the 1977 TV miniseries *Roots* as a leader of Kunta Kinte's village. He also had a pioneering role as an African-American in science fiction television. His portrayal of Lt. Travers, a member of a lunar exploration team in the *Moonstone* episode of *The Outer Limits* in 1964, pre-dated Nichelle Nichols' portrayal of a black member (Lt. Uhura) of a space exploration crew on *Star Trek*. His biggest early role was as Mike, the second male lead on the veterinary drama *Daktari*, which ran on CBS from 1966-1969. He also played Mr. MacDonald, who aided Caesar in *Conquest of the Planet of the Apes*, and in 1973 he was the star of the 'blaxploitation' film *Detroit 9000.*

Rhodes first television role was in a 1957 episode of *Zane Grey Theater* that starred Sammy Davis Jr. The role came just one year after he had had received a rude lesson in racial prejudice.

"I read about a training program a major studio had for grooming people for 'stardom.' Being naive about the system, I got on the phone and called the man in charge and asked if he would interview me, and he told me to come around to the studio," Rhodes told *TV Guide* in 1968. "I said, 'By the way, I think I should tell you that I am a Negro.' He said, 'Don't waste your time - we don't take Negroes in this program.' I hung up the phone. Almost tore the cradle off the thing."

Rhodes channeled his anger into a novel, *A Chosen Few*, which was published in a paperback edition. *A Chosen Few* was described as "an explosive personal portrait of what (Rhodes) saw and lived through in the heart of the South in the last all-Negro Marine boot camp." The novel's uneducated hero remarks, "Bitterness... is a consuming, cancerous quality out of which comes nothing but self-destruction, while out of an anger can come many constructive things, if nothing more than the drive to get something done."

Rhodes later penned two unpublished novels. *Harambee,* was about a man with a plan to liquidate the world's entire Caucasian population, and *Land of Odds* was about Hollywood.

Rhodes told TV Guide that writing served as his safety valve. "I'd rather be writing my own than reading somebody else's. I have no need for it."

Hari Rhodes died of a heart attack on January 15, 1992 in Canoga Park, California at the age of sixty-nine.

ROB RIGGLE
Daily Show Devil Dog

Robert A. "Rob" Riggle, Jr. is an actor/comedian who was born on April 21, 1970 in Louisville, Kentucky and is best known for his time as a cast member on *Saturday Night Live,* his supporting role in the film *Step Brothers*, and for his work as a correspondent on Comedy Central's *The Daily Show* from 2006-2008.

Riggle was raised in Overland Park, Kansas where he attended Shawnee Mission South High School, and he still regards it as his hometown. He is a graduate of the University of Kansas, a member of the fraternity Phi Gamma Delta, and has a Bachelor's degree in Theater and Film. He later went on to get his Master's degree in Public Administration from Webster University.

Riggle is currently a Lieutenant Colonel in the Marine Corps Reserve and has served in Liberia, Kosovo, and Afghanistan. He is currently a public affairs officer with the New York City Public Affairs unit, and is a recipient of the Combat Action Ribbon. He has referred to his military experiences on *The Daily Show*, often when acting as the show's 'Military Analyst' - and sometimes states he could kill any other member of the show. He is the only *Daily Show* cast member to have been in the military.

In August of 2007 Riggle went to Iraq to report for *The Daily Show* as well as entertain the troops under the purview of the United Service Organizations, and on April 15, 2009 it was announced that he had been promoted to the rank of Lieutenant Colonel.

He has a long-standing comedic partnership with comedian Rob Huebel with whom he frequently works at the *Upright Citizens Brigade Theater*. The duo's growing popularity in the New York comedy scene landed them an audition on *Saturday Night Live* in the summer of 2004. The two auditioned together, getting called back a number of times before the *SNL* producers finally made their choice. Only Riggle ended up making the cut. After spending one season on SNL from 2004-2005 he joined Huebel in LA to work on new projects. It was announced soon after that the two were hired by NBC in early 2006 to develop ideas for a possible new half-hour comedy program for the network.

In September of 2006 he joined the cast of *The Daily Show* to replace the departing Rob Corddry, and made his debut on September 20, 2006. Riggle left *The Daily Show* on December 10, 2008 to, in his words, "go fight crime."

In late 2007 Riggle began appearing as a spokesman in a series of Budweiser commercials, and 2008 saw him sign a talent holding contract with CBS and CBS Paramount Network TV which includes a development deal to create and star in a half-hour comedy series.

PERNELL ROBERTS
Adam Cartwright

Pernell Elvin Roberts is a television actor and singer who was born on May 18, 1928 in Waycross, Georgia. He is best known for his roles as Ben Cartwright's eldest son, Adam Cartwright, on the western series *Bonanza* (a role he played from 1959 to 1965), and as chief surgeon Dr. Trapper John MacIntyre on *Trapper John, M.D.* He is also known for his activism, which included participation in the Selma to Montgomery marches in 1965, and pressuring NBC to refrain from hiring Caucasians to portray minority characters.

During his high school years Roberts sang in local USO shows. He flunked out of Georgia Tech, and then served for two years in the Marine Corps. He then attended the University of Maryland, but also flunked out there, and began his acting career in off-Broadway and Broadway theatre in New York City. He then worked with the Arena Stage Company in Washington, D.C.

Roberts came to prominence playing Ben Cartwright's urbane eldest son, Adam, in the Western television series *Bonanza*. Despite the show's success, he left after the sixth season in 1965 due to disagreements with the writers and a desire to return to legitimate theatre. Among other complaints Roberts argued that a thirty-four-year-old, educated, Eastern-born man would not be calling his father "Pa," and the writers tacitly agreed not to exceed three "Pa" references per episode. According to producer David Dortort in the February 2006 *Bonanza Gold* issue Roberts also wanted to stop wearing his toupee. Since, in real life, there were fewer than thirteen years of age between Roberts and Lorne Greene, a bald Adam would not have translated well on screen. Bonanza continued without him for another eight seasons.

While performing in the series Roberts recorded *Come All Ye Fair and Tender Ladies*, a folk music album which *Allmusic* calls "...the softer, lyrical side of folk music - pleasant and not challenging, but quite rewarding in its unassuming way." On the *Bonanza* box set albums Roberts also sings *Early One Morning, In the Pines, The New Born King, The Bold Soldier, Mary Ann, They Call the Wind Maria, Sylvie, Lily of the West, The Water is Wide, Rake and a Ramblin Boy, A Quiet Girl, Shady Grove, Alberta,* and *Empty Pocket Blues*. Roberts was the only trained or accomplished singer of the original Ponderosa clan (David Canary, who joined the cast in 1967, graduated as a voice major), although Lorne Greene's deep baritone voice scored big in songs like 1964's *Ringo*.

Roberts continued to do guest shots on TV shows such as *The Big Valley, Mission: Impossible, The Wild Wild West,*

Gunsmoke, Mannix, The Odd Couple, Hawaii Five-O, and *The Hardy Boys.* His rich baritone voice was displayed when he played Jigger in an ABC television presentation of *Carousel,* and Rhett Butler in the Los Angeles stage production of *Scarlett.*

He regained star status in the early 1980s while starring in the television series *Trapper John, M.D.* from 1979 until 1986. Roberts ended up playing the character almost twice as long as Wayne Rogers did on the CBS series *M*A*S*H.*

He is now retired, the lone surviving original *Bonanza* cast member. Roberts made his last TV appearance on a 2001 *Diagnosis Murder,* updating a *Mannix* character he had portrayed decades before.

Roberts married three times. His first marriage was in 1951 to Vera Mowry, a professor at Washington State University with whom he had his only child, Jonathan Christopher Roberts, who died in a motorcycle accident in 1989 at age thirty-eight. Roberts and Mowry later divorced. He married Judith Anna LeBreque on October 15, 1962 and they divorced in 1971. His last marriage was to Kara Knack, whom he married in 1972. They divorced in 1996.

JOHN RUSSELL
Lawman

John Lawrence Russell (January 3, 1921 - January 19, 1991) was an actor who was born in Los Angeles, California. He was best known for playing Marshal Dan Troop in the western television series *Lawman* from 1958 to 1962.

Russell fit the Hollywood image of tall, dark, and handsome. He attended the University of California as a student athlete and following the outbreak of World War II joined the Marines, received a battlefield commission as a lieutenant at Guadalcanal, and returned home after the war as a highly decorated veteran.

He was discovered by a talent agent while in a Beverly Hills restaurant and made his film debut in 1945. Russell was contracted to 20th Century Fox in several supporting roles, and later was signed with Republic Pictures. He primarily played secondary roles, often in western films, but in 1952 starred opposite Judy Canova in *Oklahoma Annie*. In 1955 Russell was given the lead role in a television drama called

Soldiers of Fortune. The half-hour adventure show placed him and his sidekick, played by Chick Chandler, in a dangerous jungle setting. While the show proved popular with young boys, it did not draw enough adult viewers to its prime slot and was canceled in 1957.

A year later Russell was cast in his most memorable role as Marshal Dan Troop, the lead character in *Lawman*, an ABC western series that ran for five years. Co-starring Peter Brown, who played Deputy Johnny McKay, Russell played a U.S. frontier peace officer mentoring his younger compatriot.

Russell appeared in other motion pictures, notably as a supporting player in the Howard Hawks 1959 western *Rio Bravo*, which starred John Wayne, Dean Martin, and Walter Brennan. Through the 1960s to the 1980s he returned to secondary roles, and appeared in more than twenty films including three directed by his friend Clint Eastwood.

Russell appeared in the second season of the Filmation children's science fiction series *Jason of Star Command*. He played the role of Commander Stone, a blue skinned alien from Alpha Centauri. He replaced James Doohan, who had played the Commander in the previous season but left to start working on *Star Trek: The Motion Picture*.

John Russell died from emphysema in 1991 at the age of seventy and was interred in the Los Angeles National Cemetery, a former U.S. Veterans Administration cemetery, in Los Angeles, California.

ROBERT RYAN
One of the Wild Bunch

Robert Bushnell Ryan (November 11, 1909 - July 11, 1973) was an Academy Award and BAFTA award-nominated actor born in Chicago, Illinois who often played hardened cops and ruthless villains.

Ryan was born to Timothy Ryan and his wife Mabel Bushnell Ryan, and was an only child. He graduated from Dartmouth College in 1932, having held the school's heavyweight boxing title all four years of his attendance. After graduation the 6'4" Ryan found employment as a stoker on a ship, a WPA worker, and a ranch hand in Montana.

Ryan attempted to make a career in show business as a playwright, but had to turn to acting to support himself. He studied acting in Hollywood, and appeared on stage and in small film parts during the early 1940s.

In January of 1944, after securing a contract guarantee from RKO Radio Pictures, Ryan enlisted in the Marine Corps and served as a drill instructor at Camp Pendleton in California. At Camp Pendleton he befriended writer and future director Richard Brooks, whose novel *The Brick Foxhole* he greatly admired. He also took up painting.

Ryan's breakout film role was as an anti-Semitic killer in *Crossfire*, a film noir based on Brooks' novel. From then on Ryan's specialty was tough/tender roles, and he found particular expression in the films of celebrated directors such as Nicholas Ray, Robert Wise and Sam Fuller. In Ray's *On Dangerous Ground* he portrayed a burnt-out city cop finding redemption while solving a rural murder. In Wise's *The Set-Up* he played an over-the-hill boxer who is brutally punished for refusing to take a dive. Other important films were Anthony Mann's western *The Naked Spur*, Sam Fuller's uproarious Japanese set gangland thriller *House of Bamboo, Bad Day at Black Rock*, and the socially conscious heist movie *Odds Against Tomorrow.* He also appeared in several all-star war films, including *The Longest Day* and *Battle of the Bulge*.

In his later years Ryan continued playing key roles in major films. Most notable of these were *The Dirty Dozen, The Professionals,* and fellow Marine Sam Peckinpah's highly influential brutal western *The Wild Bunch.*

In contrast to the often violent and bigoted roles he played Ryan was a liberal Democrat who tirelessly supported civil rights issues, and despite his military service he also came to share the pacifist views of his wife Jessica, who was a Quaker.

During the late 1940s, as the House Committee on Un-American Activities (HUAC) intensified its anti-communist attacks on Hollywood, Ryan joined the short-lived

Committee for the First Amendment. Throughout the 1950s he donated money and services to civic and religious organizations such as the American Civil Liberties Union, American Friends Service Committee, and United World Federalists. Then in September 1959 he and Steve Allen became founding co-chairs of The Committee for a SANE Nuclear Policy's Hollywood chapter.

By the mid-1960s Ryan's political activities included efforts to fight racial discrimination. He served in the cultural division of the Committee to Defend Martin Luther King and with Bill Cosby, Robert Culp, Sidney Poitier, and other actors helped organize the short-lived Artists Help All Blacks.

Ryan's film work often ran counter to the political causes he embraced. He was a pacifist who starred in war movies, westerns, and violent thrillers. He was an opponent of McCarthyism who nevertheless served the anticommunist cause by playing a nefarious Communist agent in *I Married a Communist*. Even in films like *Crossfire* and *Odds Against Tomorrow*, which ultimately promoted racial tolerance, he played bigoted bad guys. Ryan was often vocal about this dichotomy. At a screening of *Odds Against Tom*orrow he appeared before black and foreign press representatives to discuss "the problems of an actor like me playing the kind of character that in real life he finds totally despicable."

On March 11, 1939 Ryan married Jessica Cadwalader. They had two sons, Cheyney (now a Professor of Philosophy at the University of Oregon) and Timothy "Tim" Ryan, and one daughter, Lisa. Robert and Jessica Ryan remained married until her death from cancer in 1972.

Robert Ryan died from lung cancer in New York City in 1973 at the age of sixty-three.

GEORGE C. SCOTT
Patton

George Campbell Scott (October 18, 1927 - September 22, 1999) was a stage and film actor, director, and producer who was born in Wise, Virginia, the son of Helena Agnes (née Slemp) and F. Scott. He was the only son, and the younger of their two children. His mother died just before his eighth birthday and he was raised by his father, an executive at the Buick Motor Company.

Scott was best known for his Academy Award-winning portrayal of General George S. Patton Jr. in the film *Patton*, as well as his flamboyant performance as General Buck Turgidson in Stanley Kubrick's *Dr. Strangelove*.

As a young man Scott's original ambition was to be a writer like his favorite author, F. Scott Fitzgerald, and while in high school he wrote many short stories - none of which were ever published. As an adult he tried on many occasions to write a novel, but was never able to complete one to his satisfaction. When asked by an interviewer in later life which contemporary novelists he admired he replied, "I stopped reading novels when I stopped trying to write them."

Scott joined the Marine Corps, serving from 1945 until 1949, and was assigned to the prestigious 8th and I Barracks in Washington, D.C. His duties there included serving as a guard at Arlington National Cemetery and teaching English literature and radio speaking/writing at the Marine Corps Institute. He later said his duties at Arlington, which included serving on burial details, led to his drinking.

After his military service Scott enrolled in the University of Missouri where he majored in journalism and became interested in drama. He left college after a year to pursue acting.

Scott first rose to prominence for his work with the New York Shakespeare Festival in 1958, winning an Obie Award for his performances in *Children of Paradise, As You Like It*, and an especially acclaimed performance as Shakespeare's *Richard III* (a performance one critic said was the "angriest" *Richard III* of all time). He graduated to Broadway the following year, winning critical acclaim for his portrayal of the prosecutor in *The Andersonville Trial* by Saul Levitt. This was based on the military trial of the commandant of the infamous Civil War prison camp in Andersonville, Georgia. His performance earned him a mention in *Time* magazine as a rising young actor of great intensity. In 1970 Scott directed a highly acclaimed television version of this same play. It starred William Shatner, Richard Basehart and Jack Cassidy, who was nominated for an Emmy award for his performance as the defense lawyer in this production.

In 1963 Scott was top-billed in the critically acclaimed CBS hour-long drama series *East Side, West Side* in which he and co-star Cicely Tyson played urban social workers.

Perhaps too gritty and stark for 1963, the show lasted only one season.

Scott later won wide public recognition in the film *Anatomy of a Murder*, in which he played a wily prosecutor opposite Jimmy Stewart as the defense attorney. Scott was nominated for an Academy Award for best supporting actor, and when he was notified of the nomination he called the Academy Awards a "meat parade" or "meat race."

Scott's most famous early role was in *Dr. Strangelove,* where he played the part of General "Buck" Turgidson. It was revealed on the DVD documentary that after having shot many takes of any given scene, Stanley Kubrick would frequently ask Scott to redo it in an "over the top" fashion. Kubrick would then proceed to use this version in the final cut, which Scott supposedly resented. However Kubrick did earn Scott's respect on this film, since by that time Scott was an accomplished chess player. The cast and crew noted they would often play chess between takes, and Kubrick was the only person who could routinely beat him.

Scott's portrayal of the swaggering and controversial General Patton in the 1970 film *Patton* has become, to many, his greatest performance. Many film critics and historians consider it one of the greatest performances in the history of cinema. Scott had researched extensively for the role, studying films of the general and talking to those who knew him. Having already declined an Academy Award nomination for his appearance in the 1961 film *The Hustler*, Scott returned his Oscar for *Patton*, stating in a letter to the Academy that he didn't feel himself to be in competition with other actors. However, also regarding this second rejection of the Academy Award, Scott famously said elsewhere, "The whole thing is a goddamn meat parade. I don't want any part of it." Sixteen years later, in 1986, Scott reprised his *Patton* title role in a made-for-television sequel, *The Last Days of Patton*. The movie was based on Patton's final weeks after being mortally injured in a car accident, with flashbacks of the general's life. At the time that sequel was aired Scott mentioned in a TV Guide interview that he had verbally told the Academy to donate his Oscar to the Patton Museum, but since the instructions were never put in writing it was never delivered. The Oscar is currently displayed at the Virginia Military Institute museum in Lexington, Virginia, the same institution generations of Pattons have attended. Scott did not turn down the New York Film Critics Award for his performance (of which his wife Colleen Dewhurst said, "George thinks this is the only film award worth having.")

He continued his stage work, winning Tony Award nominations for *Uncle Vanya, Death of a Salesman*, and *Inherit the Wind*, the latter despite having to miss an unusually large number of performances due to illness, with his role being taken over by National Actors Theatre artistic director Tony Randall. Scott frequently directed Broadway productions in which he did not appear, including *All God's Chillun Got Wings* and *Design for Living*, as well as being an actor/director in *Death of a Salesman, Present Laughter*, and *On Borrowed Time*.

In 1971 Scott gave two more critically acclaimed performances, as a de facto Sherlock Holmes in *They Might Be Giants*, and an alcoholic doctor in the black comedy *The Hospital*. Despite his repeated snubbing of the Academy, he was again nominated for Best Actor for the latter role. He excelled on television that year as well, appearing in an adaptation of Arthur Miller's *The Price*, an installment of the *Hallmark Hall of Fame* anthology. He was nominated for and won an Emmy Award for his role, which he accepted.

His reasoning for keeping an Emmy after rejecting an Oscar was believed to be because Emmy Award winners were chosen by blue-ribbon panels of experts, while Academy Award winners were chosen by the entire Academy membership.

The actor also starred in the popular 1980 horror film *The Changeling*, with Melvyn Douglas. He received the Canadian Genie Award for Best Foreign Film Actor for his performance. In 1981 Scott appeared alongside twenty-year-old Academy Award winning actor Timothy Hutton and newbies Sean Penn and Tom Cruise in the controversial coming of age film *Taps*. The following year he was cast as Fagin in the CBS made-for-TV adaptation of Charles Dickens' *Oliver Twist,* winning praise from audiences and critics alike for his unusual portrayal of a character which in past versions was portrayed as elderly and diminutive, the polar opposite of Scott's younger, stronger, and much more formidable version. In 1984 Scott was cast in the role of Ebenezer Scrooge in a television adaptation of *A Christmas Carol*. Critics and the public alike also praised him in this performance. Some have said his *Scrooge* ranks alongside Alastair Sim's portrayal in the 1951 theatrical film, and he was nominated for an Emmy Award for the role.

In 1990 he voiced the villain *Smoke* in the TV special *Cartoon All-Stars to the Rescue*, where his character was alongside popular cartoon characters like Bugs Bunny, and also voice-acted the villain Percival McLeach in the Disney film, *The Rescuers Down Under.* The following year he hosted the TV series *Weapons At War* on A&E TV, but was replaced after one season by Gerald McRaney of *Major Dad* fame for the last two seasons. *Weapons at War* moved to The History Channel with Scott still being shown as host for the first season. His episodes would be replaced by Robert Conrad in 2000 after Scott's death the previous year.

Scott had a reputation for being moody and mercurial while on the set. "There is no question you get pumped up by the recognition," he once said, "Then a self-loathing sets in when you realize you're enjoying it." He saw a psychiatrist four times saying, "I kept laughing. I couldn't get serious. If it helps you, it helps you. If standing on your head on the roof helps you, it helps you... if you think so." There is a famous anecdote one of his stage co-stars, Maureen Stapleton, told the director of Neil Simon's *Plaza Suite*. "I don't know what to do, I am scared of him." The director, Mike Nichols, replied, "My dear, *everyone* is scared of George C. Scott!"

Scott was married five times to Carolyn Hughes (1951–1955), Patricia Reed (1955–1960), Canadian-born actress Colleen Dewhurst (1960–1965), Dewhurst a second time (1967- 1972) and actress Trish Van Devere on September 4, 1972. He starred in several films with Van Devere, including the supernatural thriller *The Changeling* in1980. They were estranged at the time of his death.

George C. Scott died on September 22, 1999 at the age of seventy-one from a ruptured abdominal aortic aneurysm. He was interred in the Westwood Village Memorial Park Cemetery in Westwood, California, and is buried next to Walter Matthau in an unmarked grave.

BERNARD SHAW
Boys of Baghdad

While discipline started in my home, the Marine Corps enhanced that value, which became essential in my career as a journalist." - *Bernard Shaw*

Bernard Shaw is a journalist and former news anchor for CNN who was born on May 22, 1940 in Chicago, Illinois. He attended the University of Illinois at Chicago from 1963 to 1968, and later served in the Marine Corps.

Shaw began his broadcasting career as an anchor and reporter for WNUS in Chicago. He then worked as a reporter for the Westinghouse Broadcasting Company in Chicago, moving later to Washington as White House correspondent. Shaw worked in the Washington Bureau of *CBS News* from 1971 to 1977, after which he moved to *ABC News* as Latin American correspondent and bureau chief before becoming the Capitol Hill Senior Correspondent. He then left ABC in 1980 to move to CNN as its Principal Anchor.

After signing with CNN on June 1, 1980 Shaw covered some of the biggest stories of the past decades, providing live coverage of the student demonstrations in Beijing's Tiananmen Square, the 1994 earthquake in Los Angeles, the funeral of Princess Diana, President Clinton's impeachment trial and the 2000 U.S. election.

Shaw is widely known for the question he posed to Democratic presidential candidate Michael Dukakis while moderating his second debate with George H. W. Bush during the 1988 election. Knowing Dukakis opposed the death penalty, Shaw asked if he would support an irrevocable death penalty for a man who hypothetically raped and murdered Dukakis' wife, Kitty. Dukakis responded that he would not, and some critics felt he framed his response too legalistically and logically and did not address it sufficiently on a personal level. Other critics thought the question was inflammatory and unwarranted at a presidential debate.

The former Marine may be best known, however, for making television history as one of the "Boys of Baghdad."
In January of 1991 Shaw stayed behind - with Peter Arnett and the late John Holliman - after other Western reporters had deserted the city. As bombs rained down outside their hotel window the three, reporting by phone, coolly brought those images into living rooms across the world during the first attacks of the Persian Gulf War. At one point he found shelter under a desk as he reported cruise missiles flying past his window. While describing the situation in Baghdad he famously stated, "Clearly I've never been there, but this feels like we're in the center of hell."

"All kinds of ordnance was being dropped, all kinds of bombs, and I made my peace with myself that I could die at any moment," Shaw told CNN recently. "We knew the dangers around us. I always believed that two major forces - one of them supreme - saved us that night: God, and some extremely well trained and well disciplined American pilots."

In 2001, at the age of sixty, Bernard Shaw decided to retire from CNN - and the Marine Corps honored the career of the former Marine turned award-winning journalist during the Sunset Parade at the Marine Corps War Memorial. "I am surprised and pleased to be honored by the Marine Corps," stated Shaw. "While discipline started in my home, the Marine Corps enhanced that value, which became essential in my career as a journalist."

As the parade's honored guest, Shaw joined fellow Marines at the War Memorial to pay tribute to all Marines who have served their nation, past and present. "Bernard Shaw served his country with great integrity as a Marine, a model citizen, and certainly as an accomplished journalist," noted Brigadier General Richard T. Tryon, Commanding General, Marine Corps Recruiting Command. "His career embodied our core values of honor, courage, and commitment, and the Marine Corps proudly salutes him."

Bernard Shaw is married to Linda Allston, with whom he has a son and daughter. The many historic events he witnessed firsthand during his career would fill a book - and that is exactly what he is now working on. Besides an autobiography, Shaw has said that he wants to write fiction, a book of essays, and a journalism primer.

BO SVENSON
Big Swede

Boris Lee Holder Svenson is an actor who was born in Goeteborg, Sweden on February 13, 1941 and is best known for his roles in American genre films of the 1970s and 1980s.

Svenson is the son of Lola Iris Viola (née Johansson), a big band leader, actress and singer, and Birger Ragnar Svenson, a personal driver, athlete and bodyguard for the King of Sweden who fought on the side of the vastly outnumbered Finns in the Finnish-Russian Winter War from 1939 until 1940.

After emigrating to the United States in 1958 as a teenager Svenson served in the Marine Corps from 1959 to1965. After the Marines he attended UCLA and pursued a Ph.D. in metaphysics.

In the mid-'70s he took over the role of lawman Buford Pusser from Joe Don Baker in both sequels to the hit 1973 film *Walking Tall*. While the sequels were less successful than the original, he reprised the role again for the short-lived 1981 television series of the same name - a show for which he was the highest paid television actor at the time. Svenson also played the sadistic Soviet agent Ivan in the Magnum, P.I. episode *Did You See the Sunrise?*, which many fans consider to be one of the long-running show's best episodes.

An accomplished athlete, Mr. Svenson has competed in world championships, Olympic trials, and/or international competitions in judo, yachting, and track-and-field. He holds black belts in judo and aikido, and is a licensed NASCAR driver.

BOBBY TROUP
(Get Your Kicks On) Route 66!

Robert William "Bobby" Troup Jr. (October 18, 1918 - February 7, 1999) was an actor, jazz pianist and songwriter who was born in Harrisburg, Pennsylvania. He is best known for writing the popular standard *(Get Your Kicks On) Route 66,* and for his role as Dr. Joe Early on the 1970s TV series *Emergency!.*

Troup graduated from the Wharton School at the University of Pennsylvania, where he was a member of the Sigma Alpha Epsilon (ΣAE) fraternity and the Mask and Wig Club.

His earliest musical success came with the song *Daddy,* which was a regional hit in 1941. When Sammy Kaye and his Orchestra recorded *Daddy* it was number one for eight weeks on the *Billboard* Best Seller chart and the number five record of 1941. Glenn Miller and his Orchestra also performed the song on their radio broadcasts.

Troup served as a Captain in the Marine Corps during World War II. He was the first white officer to be given command of an all black unit in Jacksonville, North Carolina, where the men were living in tents and had filthy latrine conditions. At that time in Jacksonville a black person was expected to cross the street or stand in the gutter when a white person walked by. Troup's black Marines built Quonset huts, new latrines, a nightclub, a boxing ring, a basketball court and formed a basketball team, a jazz band, and an orchestra. He even had a miniature golf course installed for his men. Soon white Marines from other units began spending time in that part of camp.

Troup's light and humorous musical style was similar to that of the Nat King Cole Trio, and in the 1940s Cole had a hit with Troup's best known song *Route 66* - which became a hit for Cole and a popular standard. Chuck Berry recorded *Route 66* in 1961, and in 1964 it was one of the earliest recordings by the British rock group The Rolling Stones.

Troup produced torch singer Julie London's million selling hit record *Cry Me a River* in 1955 and they married five years later following London's divorce from actor Jack Webb, who was then directing and starring in the now-classic *Dragnet* TV show.

While he relied on songwriting royalties Troup also worked as an actor, playing musician Tommy Dorsey in the film *The Gene Krupa Story* in 1959. Later he had a memorable cameo as a disgruntled staff sergeant assigned to drive Hawkeye and Trapper John around in Japan in Robert Altman's 1970 masterpiece *M*A*S*H*. His only line of dialogue is the repeated exasperation, "Goddamn Army!" which was later modified to "Goddamn army jeep!" In 1972 Jack Webb, who had previously used Troup in a 1967 episode of *Dragnet*, cast him opposite Webb's ex-wife Julie

London in the TV series *Emergency!*. As an interesting side-note, Webb had once played a Marine drill instructor in the movie *The DI*.

London and Troup had remained on cordial terms with Webb, who had used Troup and his daughter Ronny in episodes of *Adam-12* as well as the revived *Dragnet*. In the role of Dr. Joe Early, Troup projected a relaxed amiability that brought humor to the show and contrasted with the intensity of actor Robert Fuller in the role of Dr. Kelly Brackett.

Bobby Troup died in February of 1999 at UCLA Medical Center of a massive heart attack at the age of eighty. He was buried at Forest Lawn Memorial Park in Hollywood Hills and his wife, Julie London, who died the following year, was placed in the tomb next to his.

JAMES WHITMORE
"They Call Me Mac..."

James Whitmore was born on October 1, 1921 just outside New York City in White Plains, the son of James Allen Whitmore and wife Florence Crane. The stalwart actor, particularly well known for his crime yarns and war action pictures, was Educated at Choate School in Wallingford, Connecticut and received his BA from Yale University in 1944 before serving with the Marines during World War II. Following his discharge he prepared for the stage under the G.I. bill at the American Theatre Wing, where he met first wife Nancy Mygatt. They went on to have three children together, including actor James Whitmore Jr.

Applause and subsequent kudos came quite swiftly for Whitmore both on Broadway and in film. After appearing with the Peterborough, New Hampshire Players in the summer of 1947 in the play *The Milky Way*, Whitmore made an auspicious Broadway debut as Tech Sergeant Evans in

Command Decision later that year. His gritty performance reaped the Tony, Donaldson and Theatre World awards in one fell swoop.

While the attention he merited on Broadway had Hollywood's ears perking up, the film version of *Command Decision* (starring Clark Gable) was filmed the following year - but without Whitmore repeating his stage triumph. Song-and-dance star Van Johnson, who was looking for straight, serious roles after a vastly successful musical career, was given the coveted part. The disappointment didn't last long, however. Whitmore made his film bow in 1949 with a prime role in the documentary-styled crime thriller *The Undercover Man* alongside fellow Marine Glenn Ford, and soon after made his second appearance in the war picture *Battleground*. Following its release Whitmore was the talk of the town once again at awards time - this time in Hollywood. Grabbing the Golden Globe and an Oscar nomination for 'supporting actor' as a result of his efforts, he found solid footing in the cinematic firmament. Hardly the handsome matinée lead type, he nevertheless primed himself for leading man success. Whitmore's talent, charisma and fortitude earned him a number of starring roles as well as top supporting parts in quality pictures. Gruff on the edges with a softer inner core, he appeared opposite future First Lady Nancy Davis (Reagan) in the inspirational drama *The Next Voice You Hear* as a religious, morally-minded family man, showed off his saltier side in *Mrs. O'Malley and Mr. Malone,* ably portrayed a pathetic crook in *The Asphalt Jungle* and a level-minded security chief in the stoic military drama *Above and Beyond* with Robert Taylor, played it strictly for laughs as a 'Runyonesque' gangster in the classic MGM musical *Kiss Me Kate,* and portrayed fellow Marine Tyrone Power's manager in 1956's *The Eddy Duchin Story.*

One role which is of particular interest to Marines is that of "Mac," the battle hardened sergeant who let his radio squad through the Pacific in the 1955 screen version of fellow Marine Leon Uris' novel *Battle Cry.*

Divorced from wife Nancy after two decades, he married actress Audra Lindley (Mrs. Roper of *Three's Company* TV fame) in 1972. The couple became a formidable acting pair, particularly on stage, and continued their professional partnership long after their 1979 divorce.

In the 1970s Whitmore became a magnificent one-man acting machine, bringing to life onstage such notables as Will Rogers, Harry Truman and Theodore Roosevelt. He disappeared into these historical legends so efficiently that even the powers-that-be had the good sense to preserve them on film and TV in the form of *Will Rogers' USA* (TV), *Give 'em Hell, Harry!* (which earned him his second Oscar nomination), and *Bully: An Adventure with Teddy Roosevelt.*

Whitmore has continued to earn distinction across the boards on stage, film and TV. More recently he showed Oscar potential once again with his touching role as an aged, ill-fated prison parolee in *The Shawshank Redemption*, and earned an Emmy for a recurring part on *The Practice* in the late 90s.

Currently Whitmore his enjoying his twilight years with his third wife, former actress-turned-author Noreen Nash, whom he married in 2001 as he neared the age of eighty.

LARRY WILCOX
CHiPs

"Character and integrity should be cornerstones in your life." - *Larry Wilcox*

Larry Wilcox is an actor who was born on August 8, 1947 in San Diego, California. He is best known for his role as Officer Jonathan "Jon" Baker in *CHiPs*, a popular television series about the motorcycle officers of the California Highway Patrol.

Wilcox's parents divorced while he was still an infant, and as a result he moved to Rawlins, Wyoming and was raised on *The Flying Diamond*, his grandfather's ranch. His father, a bartender, died shortly after, and his mother worked as a secretary to support her four children.

After graduating from Rawlins High School Wilcox traveled to Los Angeles and worked odd jobs while studying piano and acting. About to be drafted in 1967, he joined the Marine Corps and spent thirteen months in Vietnam as an artilleryman, reaching the rank of Sergeant before being discharged in 1970. He returned to Los Angeles to resume his acting lessons, and an attempt at pre-dentistry courses at a local college convinced him he "didn't have the manual dexterity for it."

Before long Wilcox was landing lucrative television commercials for products such as Old Spice, and in 1972 he was cast as Dale Mitchell in another reincarnation of the popular series *Lassie*, which lasted until 1974. He went on to make several guest appearances on TV, such as on *M*A*S*H*, until he was cast as Jon Baker on CHiPs in 1977. Wilcox admits taking the role to ease his despair at losing a part in *Rich Man, Poor Man* to Nick Nolte.

An accomplished motorcycle rider, horseman, race car driver and jet skier, Wilcox did many of his own stunts on the show. Unlike co-star Erik Estrada (who played Francis "Ponch" Poncherello), Wilcox never suffered any major injuries, and by the 1979–80 season he was making $25,000 per episode (the same amount as Estrada). During their time on *CHiPs* Wilcox and Estrada appeared on the cover of *TV Guide* three times.

Rumors of friction between the two surfaced during the 1978-79 season, but calmed down after Estrada's injury the following year. Apparently the accident was not enough to bring them back together however, since Estrada was not invited to his co-star's 1980 wedding. Wilcox admits he and Estrada were never the best of friends, and that some of the rumors of on-set feuding were true. Even so, whatever friction may have existed between the two never showed in Jon and Ponch's on-screen relationship. Said Wilcox, "We are just two totally different people."

In 1982, Wilcox left *CHiPs* to pursue other projects. He formed his own production company, Wilcox Productions, which produced the acclaimed television series *The Ray Bradbury Theater* for five years. He also continued acting and directing.

In the mid-1990s Wilcox ran a pharmaceutical company called *Team Elite*. He described it as "a multidivisional company selling liquid vitamins, wholesale travel and long-distance communications through network marketing."

Today he continues to run *Wilcox Productions* and is also president and chief executive officer of *UC Hub Group, Inc.*, a software company specializing in digital communities. He has also produced, directed, written, and starred in a children's video called *Lil' CHP*, which tells the story of two little boys who dream of becoming California Highway Patrol motor officers. Wilcox returned in his *CHiPs* role of Jon Baker, and former co-star Robert Pine returned as Sergeant Joe Getraer. The video also starred John Schneider (*The Dukes of Hazzard*) and Ron Masak (*Murder She Wrote*).

Wilcox says his entrepreneurial talents were nurtured during his days on *CHiPs* by David Begelman, then the top executive at MGM, which produced the show. Contrary to Estrada, who blames Begelman for sabotaging his career, Wilcox says, "I loved Begelman. He helped me so much. He gave me advice, we discussed deals and how to negotiate. He was kind of a mentor. He sure was nice to me."

With encouragement from Begelman, Wilcox optioned the rights to several entertainment properties including the story of a young actress murdered by her boyfriend, which became *Death of a Centerfold: The Dorothy Stratten Story*. Wilcox was executive producer of the TV movie, which starred Jamie Lee Curtis.

The story had a deeper resonance for Wilcox, whose older sister was shot in front of her three children and her mother.

The accused murderer, her husband, was later acquitted in a celebrated trial in Wyoming and was subsequently killed in a barroom brawl.

Wilcox was reunited briefly on-screen with his former co-star Estrada in *National Lampoon's Loaded Weapon 1*, and then again in 1998's *CHiPs* reunion movie, "*CHiPs 99*," and says he still talks to Estrada a few times a year.

Wilcox married his first wife, Judy Vagner, on March 29, 1969 and they divorced in 1978. They had two children together. Derek, the elder, was born in 1970 and appeared in two episodes of *CHiPs*. He is currently getting his Ph.D. in Intellectual History and Critical Thinking at the University of Chicago. Heidi, born in 1975, is working in the dental profession.

Wilcox's second marriage was to Hannie Strasser, a onetime *CHiPs* assistant sound technician and a native of Denmark. The wedding took place on April 11, 1980 and their daughter Wendy was born in 1982.

On April 22, 1986 Wilcox married his third wife, Marlene Harmon Wilcox, who was a member of the 1980 Olympic Heptathlon Team. Their two children are Chad and Ryan.

As of 2006 Wilcox lives on a ranch in the San Fernando Valley and is content running his company. He also states that "(I am) enjoying my family and planting as many memories as I can." He also says he remains grateful for his many fans. When asked what he thought about the upcoming *CHiPs* feature film, he simply said, "I wish them well!"

STEVE WILKOS
Once a Marine

Steven John Wilkos is a television personality who was born on March 9, 1964 in Chicago to Stanley Wilkos, a retired Chicago police officer, and Jeanette, a beauty school instructor. He currently hosts his own talk show, *The Steve Wilkos Show*, and is perhaps best known as the former director of security on *The Jerry Springer Show*.

Wilkos attended the John J. Audubon School and Lane Tech High School in Chicago. In junior high was a puny one hundred and thirty-five pounds and wore glasses, so in high school he started lifting weights and gained one hundred and ten pounds. The newly bulked up Wilkos started playing baseball his freshman year, and graduated from high school in 1982. After high school he didn't care about going to college, so he joined the Marine Corps and was enlisted for seven years until receiving an honorable discharge in 1989. The next year Wilkos followed in his father's footsteps and became a Chicago police officer. Then in May of 1994 the producers for the *Jerry Springer Show* needed security guards for a Klan show and decided to hire off-duty cops. Wilkos recruited five fellow officers, and has been with the show ever since.

"My big role model was Ted Williams, not only because he was a baseball player, but also because he was a jet fighter pilot in the Marines. He served in World War II and the Korean War while still an active baseball player for the Boston Red Sox. My other role model is my dad. Without him I wouldn't be where I am today. He also served in the Korean War, and was a Police Officer for thirty years."

Wilkos is a national spokesperson for the military charity USA Cares, and travels the country to get out the word. USA Cares was created to help bear the burdens of service by providing military families with financial and advocacy support in their time of need. The charity assists wounded warriors and their families, helps to prevent home foreclosures and evictions, and supports veterans and their families during times of financial hardship.

Wilkos has a special place in his heart for those in the service, both the veterans themselves and the families left behind by those serving overseas. "As a former Marine, I couldn't be more proud to support an organization like USA Cares," said Wilkos. "USA Cares provides military families and veterans with critical resources and support, and I cannot think of a more deserving group of people."

Bill Nelson, Executive Director of USA Cares said, "Steve Wilkos takes on difficult challenges and issues in a very direct and impactful way. It's like my Marine Corps friends say, 'once a Marine ... always a Marine.' You have to respect the fact that some challenges just cannot be ignored, and USA Cares is appreciative of Steve's commitment to assisting us help Military Families and Veterans from all branches of service when they most need it."

Steve Wilkos' third wife, Rachelle Consiglio, is the executive producer for *Springer* as well as the *Steve Wilkos Show*. They have two children and live in Park Ridge, Illinois.

MONTEL WILLIAMS
The Montel Williams Show

Montel Brian Anthony Williams is a television and radio talk show host who was born on July 3, 1956 in Baltimore Maryland. In high school he was an outstanding student, athlete, musician and student body president. His father, Herman Williams, Jr., was a firefighter who became the first African-American Fire Chief in Baltimore in 1992.

Williams enlisted in the Marine Corps in 1974 and completed recruit training at MCRD Parris Island, South Carolina. Afterwards, while training at Twentynine Palms, he was selected for training at the Naval Academy Preparatory School and a year later he was accepted into the United States Naval Academy. He graduated in 1980 with a degree in engineering and a minor in international security affairs and was commissioned as an Ensign in the United States Navy.

Williams served aboard *USS Sampson* during the U.S. invasion of Grenada in 1983, and after twelve years of military service departed the Navy as a Lieutenant Commander.

In 1991 he was a rodeo clown in Switzerland for a year prior to going on TV as a talk show host on the *The Montel Williams Show*. Then in 1996 Williams received a Daytime Emmy Award for Outstanding Talk Show Host. He was again nominated for Outstanding Talk Show Host in 2002, and the *Montel Williams Show* was nominated for Outstanding Talk Show in 2001 and 2002.

Williams also guest-starred in episodic television and off-Broadway plays. Among others, he portrayed a Navy SEAL lieutenant in several episodes of the television series *JAG*. Williams also produced and starred in a short-lived television series called *Matt Waters*, which appeared on CBS in 1996, in which he played an ex-Navy SEAL turned inner-city high school teacher.

Williams also played the judge presiding over Erica Kane's (Susan Lucci) murder trial on the ABC soap opera *All My Children* in 2002. Then in 2003 he made a guest appearance on the soap as himself to promote an episode of his own show on which several *AMC* stars were scheduled to appear. In 2004 he hosted *American Candidate*, a political reality show for Showtime.

On January 30, 2008 *Variety* reported that CBS TV Distribution had terminated *The Montel Williams Show* when key Fox-owned stations chose not to renew it for the 2008-2009 season. This followed an appearance on the show *Fox & Friends* in which he criticized the media's lack of coverage on the Iraq War and took the hosts to task for their (and the media in general) excessive coverage of the death of actor Heath Ledger, contrasted with the sparse coverage of U.S. troops dying in Iraq. The last episode of *The Montel Williams Show* aired on May 16, 2008.

Williams has two daughters, Ashley and Maressa, with his first wife, Rochele See. He then married Grace Morley, a burlesque dancer, on June 6, 1992. They had a son, Montel Brian Hank, and a daughter, Wynter Grace, but divorced in 2000. In July of 2006 Williams proposed to girlfriend Tara Fowler, an American Airlines flight attendant, and they married before friends and family on a beach in Bermuda on October 6, 2007.

Williams participated in the 2007 World Series of Poker main event and planned to donate any potential winnings to families affected by the Iraq war, but was eliminated on Day Two.

Williams was diagnosed with Multiple Sclerosis in 1999 and now heads the Montel Williams MS Foundation. He is also a national spokesman of the Partnership for Prescription Assistance, a patient-assistance program clearinghouse that helps low-income patients apply for free or reduced-priced prescription drugs.

JONATHAN WINTERS
Gunny Davis

"I've always been proud of being a Marine. I won't hesitate to defend the Corps." - *Jonathan Winters*

Jonathan Harshman Winters III is a comedian and actor who was born in Bellbrook, Ohio on November 11, 1925. The son of radio personality Alice Kilgore (née Rodgers) and Jonathan Harshman Winters II, an investment broker, he is a descendant of Valentine Winters, founder of the Winters National Bank in Dayton, Ohio (now part of JPMorgan Chase). When he was seven his parents separated, and Winters' mother took him to Springfield, Ohio to live with his maternal grandmother.

At the age of seventeen Winters joined the Marine Corps and served two and a half years in the Pacific Theater during World War II. Upon his return he attended Kenyon College in Gambier, Ohio and later studied cartooning at Dayton Art

Institute where he met Eileen Schauder, whom he married in 1948.

He began comedy routines and acting while studying at Kenyon College, and was also a local radio personality on WING in Dayton and WIZE in Springfield. He then performed as Johnny Winters on WBNS-TV in Columbus for two years, quitting the station in 1953 when they refused him a five dollar raise. After promising his wife that he would return to Dayton if he did not make it in a year, and with just fifty-six dollars in his pocket, he moved to New York City and stayed with friends in Greenwich Village. After obtaining Martin Goodman as his agent, he began stand-up routines in various New York nightclubs. His big break occurred (with the revised name of Jonathan) when he worked for Alistair Cooke on the CBS Sunday morning show *Omnibus*. In 1957 he performed in the first color television show, a fifteen-minute routine sponsored by Tums.

As a stand-up comic with a madcap wildness, Winters recorded many classic comedy albums for the Verve Records label starting in 1960. Probably the best-known of his characters from this period is Maude Frickert, the seemingly sweet old lady with a barbed tongue. He was a favorite of Jack Paar, and appeared frequently on his television programs. In addition he would often appear on the *The Tonight Show Starring Johnny Carson*, usually in the guise of some character. Carson often did not know what Winters had planned, and usually had to tease out the character's back story during a pretend interview.

Winters has appeared in nearly fifty movies and several television shows, including particularly notable roles in the film *It's a Mad, Mad, Mad, Mad World* and in the dual roles of Henry Glenworthy and his dark, scheming brother, the Reverend Wilbur Glenworthy, in the film adaptation of

Evelyn Waugh's *The Loved One*. Fellow comedians who starred with him in *Mad World*, such as Arnold Stang, claimed that in the long periods while they waited between scenes Winters would entertain them for hours in their trailer by becoming any character they would suggest.

On television in the late sixties he appeared as a regular (along with Woody Allen and Jo Anne Worley) on the Saturday morning children's program *Hot Dog*, and in the seventies he appeared in his own show, *The Wacky World of Jonathan Winters*.

In 1991 and 1992 he was on *Davis Rules*, a sitcom that lasted two seasons. He played 'Gunny' Davis, an eccentric retired Marine who was helping raise his grandchildren after his son had lost his wife

In 1999 Winters was awarded the Mark Twain Prize for American Humor, and in June of 2008 he was presented with the TV Land Pioneer Award by his friend Robin Williams.

On January 11, 2009 Eileen, Jonathan Winters's wife of sixty years, died at the age of eighty-four after a twenty-year battle with breast cancer. He now lives near Santa Barbara, California and is often seen browsing and hamming for the crowd at the antique show on the Ventura County fairgrounds and entertaining the tellers and other workers whenever he visits his local bank to make a deposit or withdrawal. He spends time painting, and has been presented in one-man shows of his art. In 1987 he published *Winters' Tales: Stories and Observations for the Unusual*. Other writings have followed, and he is said to be working on his autobiography.

BURT YOUNG
"Yo, Paulie!"

Burt Young is an Academy Award-nominated actor, painter, and author who was born Jerry De Louise in Queens, New York on April 30, 1940, the son of Josephine and Michael Young.

Young began acting after serving in the Marines in the late 1950s, and also worked as a carpet salesman and amateur boxer. Young was trained by Lee Strasberg at the world-famous Actor's Studio and made his name playing rough-edged working class Italian-American characters. The best-known example was his signature role as Sylvester Stallone's brother-in-law 'Paulie' in *Rocky*, for which he received an Oscar nomination for Best Supporting Actor. He is one of only three actors (Stallone and Tony Burton being the other two) who have appeared in every *Rocky* film.

He has played similar roles in *Chinatown, Convoy, The Pope of Greenwich Village, Once Upon a Time in America,*

Last Exit to Brooklyn, Downtown: A Street Tale, and *Amityville II: The Possession.* Young has also appeared on many television programs, including *Baretta, Law & Order, Walker, Texas Ranger, M*A*S*H* , and an appearance on *The Sopranos* as an old mobster dying of cancer who comes out of retirement to execute a hit on a godson whom he hates.

Young is a painter whose art has been displayed in galleries throughout the world and a published author whose works include two filmed screenplays and a four hundred page historically based novel called *Endings.* He has also written two stage plays: *SOS,* and *A Letter to Alicia and the New York City Government From a Man With a Bullet in His Head.*

Young first gathered notice playing tough thugs in such films as *The Gang That Couldn't Shoot Straight, Across 110th Street, Chinatown* and *The Gambler.* Fiery director and fellow Marine Sam Peckinpah cast him as the getaway driver/assassin 'Mac' in 1975's *The Killer Elite,* and Young came to the attention of newcomer Sylvester Stallone who then cast him as 'Paulie.' He has also appeared in numerous other major productions including *Mickey Blue Eyes* and *The Adventures of Pluto Nash.*

Burt Young is widowed and has one daughter, actress Anne Morea. He currently owns a restaurant in the Bronx, New York.

SPORTS

DUSTY BAKER
Hard Bake

Johnnie B. "Dusty" Baker, Jr. is a Major League Baseball outfielder and manager who was born on June 15, 1949 in Riverside, California. He is currently the manager of the Cincinnati Reds, and served in the Marine Corps Reserves from 1969 through 1975. He previously led the San Francisco Giants and Chicago Cubs, winning the 2002 National League pennant with the Giants.

Drafted by the Atlanta Braves in the 1967 amateur draft out of Del Campo High School near Sacramento, California, Baker began his professional baseball career as an outfielder for the Braves in 1968. After spending sixteen full seasons with Atlanta and the Los Angeles Dodgers, as well short tenures with both the San Francisco Giants and Oakland Athletics, Baker finished his career as a player in 1986 with .278 batting average, 242 home runs, and 1,013 runs batted in. Highlights of his eighteen seasons on the diamond

227

include playing for the National League All-Star team in 1981 and 1982, winning League Championship series in 1977, 1978, and 1981, and ultimately winning a World Series title in 1981 with the Dodgers.

Baker also earned a spot as a footnote to history. On April 8, 1974 he was in the on deck circle when Hank Aaron hit homer number 715 to pass Babe Ruth in career home runs. He then went to the plate and "hit a double that nobody saw, and nobody cared about" in his subsequent at-bat.

CARMEN BASILIO
The Onion Farmer

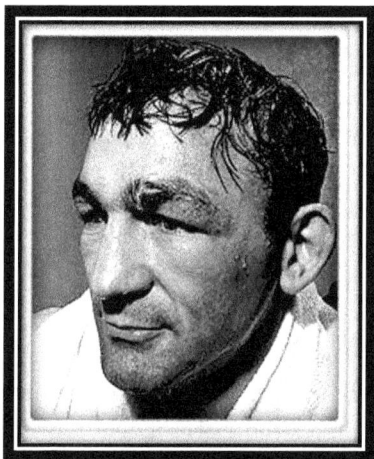

Carmine Basilio (better known in the boxing world as Carmen Basilio) is a former boxer who was born on April 2, 1927 in Canastota, New York.

After serving an enlistment in the Marine Corps, Basilio became a professional boxer in 1948. He received an opportunity to fight for a world championship in the sixth year of his professional career, but lost that welterweight title match to Kid Galivan on September 18, 1953 in a fifteen-round decision. He got a second title opportunity on June 10, 1955 and, in what has become a favorite fight of classic sports channels such as ESPN, knocked out Tony DeMarco in the twelfth round to win the welterweight championship. Basilio then had two non-title bouts, including a ten round decision win over Gil Turner, before he and DeMarco met again - this time with Basilio as the defending world champion. He won the rematch, once again in a twelve round knockout.

In his next fight, in 1956, Basilio lost the title in Chicago to Johnny Saxton by decision in fifteen rounds. It has long been speculated that the reason Saxton got the nod was he supposedly had ties with Chicago's underworld which, according to the suggestion, might have paid off the fight's judges to throw the fight to Saxton. This has been an unverified rumor which many magazines, *Ring Magazine* included, have talked about in the past. In an immediate rematch, which was boxed in Syracuse rather than Chicago, Basilio regained the crown with a nine round knockout. Then, in a rubber match, Basilio kept the belt by a knockout in two rounds.

After that he went up in weight and challenged world Middleweight champion Sugar Ray Robinson in what may have been his most famous fight. He won the championship on September 23, 1957 by beating Robinson in one of the most exciting fifteen round decisions in middleweight history. The next day he had to abandon the welterweight title, according to boxing laws of the day, and later that year Basilio won the Hickok Belt as top professional athlete of the year.

On March 25, 1958 Basilio and Robinson met in a rematch and Robinson barely regained the title with a controversial fifteen round decision. Basilio's left eye was swollen shut from the sixthround on, and yet many members of the ringside press thought Basilio won this second fight. It was a split decision, just as their first fight was.

Basilio announced his retirement from boxing in 1961, with a career record of fifty-six wins (twenty-seven by knockout) and sixteen losses, and was inducted into *Ring Magazine's* Boxing Hall of Fame in 1969, the International Boxing Hall of Fame in 1990, and the Marine Corps Sports Hall of Fame in 2002.

HANK BAUER
Old Potato Face

"One thing the Marines and Yankees have in common is called pride. Once you put those Yankee pinstripes on, you knew you had something to live up to. It's the same with the Corps. Once you get that green uniform, damn it, you knew you had to shape up. After all, they're both winners." – Hank Bauer

Henry A "Hank" Bauer was born in East St Louis, Illinois on July 31, 1922. The youngest of nine children, Bauer's father was an Austrian immigrant who worked as a bartender after losing his leg in an aluminum mill.

After graduating from Central Catholic High School, Bauer went to work repairing furnaces in a beer-bottling plant when his older brother Herman - who was playing in the White Sox farm system - was able to get him a tryout which resulted in a contract with Oshkosh Giants of the Wisconsin State League. Alternating between infield and outfield, he batted .262.

In January of 1942 Bauer enlisted in the Marine Corps. He took basic training at Mare Island, California, where he also played for the camp baseball team, but the easy life soon came to an abrupt halt.""One morning," Bauer told *Time* magazine in 1964, "this sergeant came up to me and said, 'Why don't you volunteer for the Raider battalion?' I said okay. But the first thing they told me was, 'You've got to swim a mile with a full pack on your back.' I said, 'Hell, I can't even swim!' and they turned me down. I told the sergeant what happened, and he said, 'You gutless SOB, go back down there.' So I told them I knew how to swim... and they took me."

During World War II Bauer served with G Company, 2nd Battalion, 4th Regiment, 6th Marine Division and came down with malaria almost as soon as he hit the South Pacific. "My weight dropped from 190 to 160 pounds," he said. "I was eating atabrine tablets like candy." Temporarily recovered (over the next four years, Bauer had twenty-four malarial attacks), he fought on New Georgia and was hit in the back by shrapnel on Guam. Next came Emirau off New Guinea, and then Okinawa. Sixty-four men were in Platoon Sergeant Bauer's landing group on Okinawa, and only six got out alive. Hank himself was wounded again on June 4, 1945. "I saw this reflection of sunshine on something coming down. It was an artillery shell, and it hit right behind me." A piece of shrapnel tore a jagged hole in Bauer's left thigh. Also wounded that day was Richard C. Goss, who was serving with Bauer. "There goes my baseball career," Bauer told Goss as they were evacuated together. Bauer's part in the war was over - after thirty-two months of combat, eleven campaign ribbons, two Bronze Stars, and two Purple Hearts.

His brother Herman was not so fortunate, as he was killed

in action in France with the 3rd Armored Division on July 12, 1944.

Bauer felt there was no future for him in baseball so he joined the pipe fitters' union in East St. Louis and got a job as a wrecker dismantling an old factory, but a roving baseball scout named Danny Menendez found him and offered him a tryout with the Quincy Gems, a Yankees' farm club.

Bauer hit .323 at Quincy and promptly moved up to the Kansas City Blues, where he hit .313 in 1947 and .305 in 1948. Bauer played in nineteen games with the Yankees in 1948 and one hundred-plus games in Yankees' pinstripes for each of the next eleven seasons, in addition to nine World Series appearances.

Hank broke in quickly as the Yankees' right fielder, stroking singles in his first three at-bats and proving he was there to stay. Bauer managed to get into 103 games in 1949, belting ten home runs and forty-five RBI to a .272 clip, while splitting time with Joe DiMaggio (whose playing time was limited due to injuries), Gene Woodling and Cliff Mapes. The Yankees would go on to win the World Series in five games over the Brooklyn Dodgers, the first of nine World Series that Bauer would play in with the Yanks, and the first of seven championships.

In 1950 he improved to thirteen homers, seventy RBI and a robust .320 average, good for tenth in the AL while still sharing the three outfield spots among the same three players. 1951 saw a drop in his offensive numbers (ten homers, fifty-four RBI, .296 AVG), but his rock solid defense, powerful arm and ferocious style of play were more important in a four man outfield rotation that now included a young Mickey Mantle as well as Joe DiMaggio and the consistent Gene Woodling. A Red Sox shortstop said of

Bauer, "When Hank came down the base path, the whole earth trembled!" Bauer's outlook was, "It's no fun playing if you don't make somebody else unhappy. I do *everything* hard."

With the retirement of DiMaggio after the 1951 season the Yankees decided to go with a traditional three man outfield starting in 1952, and Bauer won the full-time job in right field. He responded with his finest offensive season to date with seventeen home runs, seventy-four RBI, a .293 AVG and his first selection (of three in his career) to an all-star game. He also finished tenth in the AL in doubles, runs scored and hits), and sixth in total bases (256 - second on the team to Mantle) and extra-base hits. Defensively, he also contributed sixteen outfield assists.

His next six years were the model of consistency for Bauer. He would average about sixteen home runs, sixty RBI, seventy-nine runs scored and a .272 AVG from 1953 to 1958, with two more all-star appearances in 1953 and 1954.

Off the field, Bauer was involved in the infamous "Copacabana incident." A group of Yankee players including Bauer, Mickey Mantle, Whitey Ford, and Billy Martin, who were accompanied by their wives, became involved in a confrontation when another group of patrons at the Manhattan nightclub began to hurl racial slurs at Sammy Davis Jr., who was onstage. This infuriated Martin, whose roommate Elson Howard was the first African-American to play for the Yanks. One patron, a Bronx delicatessen owner, sued Bauer and accused him of punching him. The man lost the lawsuit after catcher Yogi Berra testified, "Nobody never hit nobody,"although various sources over time have claimed that Bauer did indeed throw and land a punch.

While there was a slight decline in his offense output after 1956, Hank's World Series play reached another level. He hit .429 in a losing effort in the 1955 Series, and a 2-for-5 performance in Game One of the 1956 World Series began a streak of hitting safely in a record seventeen straight World Series games. In his final World Series, a seven-game victory over the Milwaukee Braves, Hank went out in style, hitting at a .323 clip with four HR, eight RBI and six runs scored. In Game Two, with the Yankees down 2-0 and returning to New York, Hank drove in all four runs in a 4-0 win with a two-run single in the fifth and a two run home run in the seventh.

Once again, Hank got the Yankees going in Game Six. Down three games to two and back in Milwaukee, Hank slammed his fourth homer of the series in the top of the first inning off future Hall of Famer Warren Spahn to give the Yankees an early lead. The Yankees would go on to win a nail-biter in ten innings by a tally of 4-3, forcing a seventh game which was won by the Yankees.

Bauer would complete his Yankee career with 158 HR, 654 RBI, 703 runs scored, 1,326 hits and a .277 average in 1,406 games. After his playing days were over he managed the Kansas City Athletics and Baltimore Orioles, and in 1966 led the Orioles to the World Series where they defeated the Dodgers in four games.

Hank Bauer died of cancer in Shawnee Mission, Kansas on February 9, 2007 at the age of eighty-four.

PATTY BERG
Lady on the Links

Patricia Jane Berg (February 13, 1918 - September 10, 2006) was a professional golfer who was born in Minneapolis, Minnesota. She was a founding member of and a leading player on the Ladies Professional Golf Association (LPGA) Tour during the 1940s, 1950s and 1960s, and her fifteen major title wins remains the all-time record for most by a female golfer.

Berg took up golf in 1931 and began her amateur career in 1934, winning her first title that year - the Minneapolis City Championship. She came to national attention by reaching the final of the 1935 U.S. Women's Amateur, losing to Glenna Collett-Vare in Vare's final Amateur victory. She won the Titleholders in 1937, and then the Amateur in 1938 at Westmoreland. After winning twenty-nine amateur titles, Berg turned professional in 1940.

During World War II Patty Berg was a lieutenant in the Marine Corps from 1942 to 1945. Then in 1948 she helped establish, and became the first president of, the LPGA. She won the inaugural U.S. Women's Open in 1946, a total of fifty-seven events on the LPGA and WPGA circuit, and was runner-up at the 1957 Open at Winged Foot. In addition Berg won the 1953, 1957, and 1958 Western Opens and the 1955 and 1957 Titleholders, both considered majors at the time. She was voted the Associated Press Woman Athlete of the Year in 1938, 1942 and 1955, and her last victory came in 1962.

In 1963 she was voted the Bob Jones Award, the highest honor given by the United States Golf Association, in recognition of her distinguished sportsmanship in golf. Berg also received the 1986 Old Tom Morris Award from the Golf Course Superintendents Association of America, GCSAA's highest honor, and the LPGA established the Patty Berg Award in 1978.

Patty Berg announced in December of 2004 that she had been diagnosed with Alzheimer's disease, and she died in Fort Myers, Florida from complications of the disease twenty-one months later at the age of eighty-eight.

ROD CAREW
Sir Rodney

Rodney Cline "Rod" Carew is a former Major League Baseball player who was born on October 1, 1945 to a Panamanian mother on a train in the town of Gatún, which at that time was in the Panama Canal Zone. The train was racially segregated, with white passengers being given the better forward cars while non-whites like Carew's mother were forced to ride in the rearward cars. When she went into labor a Jewish physician traveling on the train, Dr. Rodney Cline, delivered the baby and he was named Rodney Cline Carew in appreciation.

When Rod was fourteen the Carews immigrated to the United States and lived in the Washington Heights section of the borough of Manhattan in New York City. Although Carew attended George Washington High School, which current MLB star left fielder Manny Ramirez also attended, he never played baseball for the school team. Instead he played sandlot (semi-pro) baseball for the Bronx Cavaliers, which is where he was discovered by Minnesota Twins' scout Monroe Katz. Carew signed an amateur free agent contract with the Twins a day after graduating, and three years later was called up and became a teammate of first baseman Harmon Killebrew. It was during that period in the 1960s that Carew served a six-year enlistment in the Marine Corps Reserve as a combat engineer.

Rod Carew won the American League's Rookie of the Year award in 1967 and seven batting titles over the course of his career, and was an All-Star in every season but his final one in 1985.

In 1972 Carew led the American League in batting, hitting .318, and in 1977 he won the American League's Most Valuable Player award with a batting average of .388, which at the time was the highest since fellow Marine Ted Williams hit .406 in 1941.

In 1975 Carew joined Ty Cobb as the only players to lead their leagues in batting average for three consecutive seasons, achieving this feat in 1973, 1974, and 1975. He also stole home seventeen times in his career, including seven in the 1969 season alone.

Originally a second baseman, Carew moved to first base in September of 1975. In 1979, frustrated by the Twins' inability to keep young talent, and after considerable conflict with team owner Calvin Griffith, Carew announced his intention to leave the team and was subsequently traded to the Angels for outfielder Ken Landreaux, catcher/first baseman Dave Engle, right-handed pitcher Paul Hartzell, and left-handed pitcher Brad Havens.

On August 4, 1985 Carew joined an elite group of ballplayers when he got his 3,000th base hit against Minnesota Twins left-hander Frank Viola at the former Anaheim Stadium. Coincidentally, Chicago White Sox right-hander and fellow Marine Tom Seaver won his 300th career game on the very same day. Rod Carew finished his career with 3,053 hits, and a lifetime batting average of .328.

After the season1985 season, which ended up being his last, Carew became a free agent but received no contract offers from other teams. He suspected that baseball owners were deliberately colluding to keep him from playing, and that suspicion proved to be justified. On January 10, 1995, nearly a decade after his forced retirement, arbitrator Thomas Roberts ruled that the owners had indeed violated the rules of baseball's second collusion agreement, which they had previously agreed to abide by. Carew was awarded damages of $782,036, which was equivalent to what he would have likely received in 1986.

Rod Carew was elected to the Baseball Hall of Fame in 1991 in his first year of eligibility, becoming only the 22nd player so elected. In 1999 he ranked sixty-first on *The Sporting News'* list of the One Hundred Greatest Baseball Players, and was nominated as a finalist for Major League Baseball's All-Century Team. Rod Carew was also inducted into the Hispanic Heritage Baseball Museum Hall of Fame.

ROBERTO CLEMENTE
Arriba!

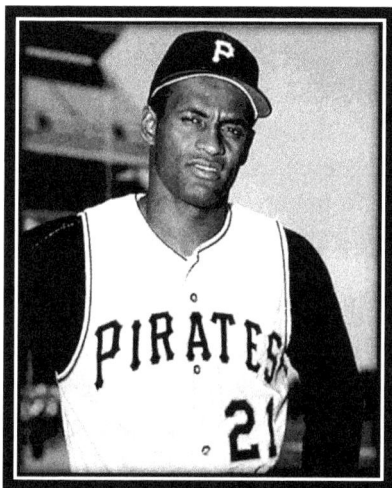

Roberto Clemente (August 18, 1934 - December 31, 1972) was a professional baseball player who was born in Carolina, Puerto Rico. He began his professional career with the Santurce Crabbers in the Puerto Rican Professional Baseball League, playing there until the Brooklyn Dodgers offered him a contract to play with the Montreal Royals. Clemente accepted the offer, and was active with the team until he was drafted by the Pittsburgh Pirates in the Major League Baseball draft on November 22, 1954.

Clemente played eighteen seasons in the Major Leagues from 1955 to 1972, all with Pittsburgh. He was awarded the National League's Most Valuable Player Award in 1966, and during the course of his career was selected to participate in the league's All Star Game on twelve occasions. He won twelve Gold Glove Awards, and led the league in batting average four different seasons. Clemente is also the first

Latino to win a World Series as a starter, a league MVP award, and be named World Series MVP.

Clemente debuted with the Pittsburgh Pirates on April 17, 1955 in the first game of a doubleheader against the Brooklyn Dodgers. At the beginning of his time with the Pirates he experienced frustration because of racial tension between himself, the local media, and even some of his teammates. Clemente responded to this by stating, "I don't believe in color." He noted that during his upbringing he was taught to never discriminate against someone else based on ethnicity.

During the middle of the season Clemente was involved in a car accident which caused him to miss several games with an injury to his lower back. He finished his rookie campaign with an average of .255, despite trouble hitting certain types of pitches

During the winter season of 1958-59 Clemente didn't play winter baseball in Puerto Rico as he usually did, but instead served in the Marine Corps Reserve for six months at Parris Island, South Carolina and Camp Lejeune, North Carolina. At Camp Lejeune he served as an infantryman, and the rigorous training program helped Clemente physically. He added strength by gaining ten pounds, and said his back troubles had virtually disappeared. He remained in the reserves until September of 1964.

Clemente spent much of his time during the off-season involved in charity work. When Managua, the capital city of Nicaragua, was affected by a massive earthquake on December 23, 1972, Clemente (who had been visiting Managua three weeks before the quake) immediately set to work arranging emergency relief flights. He soon learned, however, that the aid packages on the first three flights had been diverted by corrupt officials of the Somoza government and had never reached victims of the quake.

Clemente decided to accompany the fourth relief flight on December 31, 1972, hoping his presence would ensure the aid would be delivered to the survivors. Unfortunately the airplane he chartered for a New Year's Eve flight, a Douglas DC-7, had a history of mechanical problems and sub-par flight personnel, was overloaded by five-thousand pounds, and crashed into the ocean off the coast of Isla Verde, Puerto Rico immediately after takeoff. A few days after the crash the body of the pilot and part of the fuselage of the plane were found. An empty flight case apparently belonging to Clemente was the only personal item recovered. Clemente's teammate and close friend Manny Sanguillen was the only member of the Pirates not to attend Roberto's funeral. The catcher chose instead to dive into the waters where Clemente's plane had crashed in an effort to find his teammate, but in the end his body was never recovered.

Roberto Clemente was elected to the Baseball Hall of Fame posthumously in 1973, thus becoming the first Latin American to be selected, and the only current Hall of Famer for whom the mandatory five year waiting period was waived since the rule was instituted in 1954.

JERRY COLEMAN
The Colonel

Gerald Francis "Jerry" Coleman is a former Major League Baseball second baseman and later a play-by-play radio announcer for the San Diego Padres who was born on September 14, 1924 in San Jose, California. He spent his entire playing career with the New York Yankees and played six years in their minor league system before reaching the big club.

Coleman joined the Marine Corps on October 23, 1942 as a naval aviation cadet in the V-5 program in San Francisco, California. After going through pilot training in Colorado, Texas, and North Carolina he was commissioned a Second Lieutenant in the Marine Corps. He received his wings of gold, signifying he was a naval aviator, on April 1, 1944 at Naval Air Station, Corpus Christi, Texas and was assigned to Naval Air Station, Jacksonville, Florida where he was trained to fly the Douglas SBD Dauntless Dive Bomber. He was briefly stationed at the Marine Corps Air Station at

Cherry Point, North Carolina and was then transferred to Marine Corps Air Station El Toro, California before boarding a troop ship and being sent to Guadalcanal in the Solomon Islands as a replacement pilot.

Coleman arrived at Guadalcanal in August pf 1944 and was assigned to VMSB-341, known as "The Torrid Turtles." He flew fifty-seven combat missions, flying close air support - which VMSB-341 was the first squadron in the Marine Corps specifically designated to do - and also flew missions in the Solomon Islands and the Philippines. In July of 1945 his squadron, along with other Marine Corps squadrons, were called back from the Pacific to form carrier-based groups in anticipation of the amphibious assault on Japan. When the war ended he remained stationed at Cherry Point, and in January of 1946 he was transferred from active duty to the inactive reserve list and resumed his baseball career.

In 1949 Jerry Coleman became a member of the New York Yankees. He played second base and was third in Rookie of the Year balloting with a batting average of .275, two home runs, and forty-two runs batted in. The following season he was an American League All-Star, and was the Most Valuable Player in the 1950 World Series when the Yankees swept the Philadelphia Phillies in four games. He played in a total of six World Series with the Yankees, winning four. Coleman is regarded as one of the best defensive second basemen of all time, having committed only 89 errors in 3,168 fielding opportunities and turning 532 double plays.

With the outbreak of the Korean War Coleman was recalled to active duty and sent to El Toro for training in the Vought F4U Corsair. He was assigned to VMF-323, also known as the "Death Rattlers," and flew sixty-three combat missions in the F4U and AU1 Corsair. This included close

245

air support and interdiction/strike missions. He was then assigned duties as a forward air controller. Coleman was transferred back to the United States in August of 1953, and was again placed on the reserve list. Later that month he was back playing second base for the New York Yankees, and that same year he joined a Marine Corps Reserve unit in New York where he did promotional work for the Corps until he retired as a Lieutenant Colonel in 1964. Having served three years in World War II and two years in Korea, he is the only major league player to see combat in two wars and was awarded two Distinguished Flying Crosses, thirteen Air Medals, and three Navy Citations.

After retiring from baseball Coleman began his distinguished career as a sports announcer with the New York Yankees (1963 - 1969), the California Angels (1970 - 1971), and CBS Radio's Network Game of the Week (for twenty-two seasons). He is currently the radio voice of the San Diego Padres (1972 - 1979, 1981 - present), and is famous for the phrases, "Oh, doctor!" and "You can hang a star on that!"

Jerry Coleman was inducted into the Padres Hall of Fame in 2001, was named winner of the Ford C. Frick Award for broadcasting excellence in 2005, and was inducted into the Baseball Hall of Fame in Cooperstown, New York.

EDDIE COLLINS
Cocky

"Outside of my accomplishments on the baseball field, I am most proud of my one-time membership in the greatest of all military units - the United States Marines."
– *Eddie Collins*

Edward Trowbridge Collins, Sr. (May 2, 1887- March 25, 1951), nicknamed "Cocky," was a Major League Baseball second baseman, manager and executive who played from 1906 to 1930 for the Philadelphia Athletics and Chicago White Sox. He was a star on the Athletics' "$100,000 infield" which propelled the team to four American League pennants and three World Series titles between 1910 and 1914, and was named the league's Most Valuable Player in 1914. After that season his contract was sold to the White Sox, and he helped them capture pennants in 1917 and 1919. At the end of his career he ranked second in major league history in career games (2,826), walks (1,499) and stolen bases (744), and was third in runs scored (1,821), fourth in hits (3,315) and at bats (9,949), sixth in on base percentage (.424), and eighth in total bases (4,268). He was also fourth in AL history in triples (187), and still holds the major league record of 512 career sacrifice hits - over one hundred more than any other player. He was the first major leaguer in modern history to steal eighty bases in a season, and still shares the major league record of six steals in a game, which he accomplished twice in September of 1912. He regularly batted over .320, retiring with a career average of .333. He also holds major league records for career games (2,650), assists (7,630) and total chances (14,591) at second base, and ranks second in putouts (6,526). Under the win shares statistical rating system created by baseball historian and analyst Bill James, Collins was the greatest second baseman of all time.

Eddie Collins was one of the most accomplished all-around ballplayers ever to play the game. They called him "Cocky," not because he was arrogant, but because he was filled with confidence based on sheer ability. Bill James wrote, "Collins sustained a remarkable level of performance for a remarkably long time. He was past thirty when the lively ball era began, yet he adapted to it and continued to be one of the best players in baseball every year... his was the most valuable career that any second baseman ever had."

Collins played for twenty-five years, twenty of them as a regular. He won no batting titles because he played during the same era as Ty Cobb, but did lead the AL in stolen bases four times and in runs scored three consecutive seasons, 1912-14. He was a superlative fielder, leading all second basemen in fielding average nine times, and a brilliant base runner. In the dugout or on the coaching lines, he was a canny, sign-stealing, intuitive strategist.

Collins' background was atypical of a player of the early 1900s. He starred as captain of Columbia University's baseball team, and although he was barred from playing his senior year because he had disguised himself as "Eddie Sullivan" to play professionally (even getting into a few games with the Athletics), was named Columbia's coach and stayed to get his degree.

Collins was one of the key young players on Connie Mack's great Athletic teams of 1909-14 that won pennants in all but 1912. When Mack disbanded his long-reigning team in 1915 he sold Collins to the Chicago White Stockings. Collins starred in the 1917 World Series, hitting .409, and scoring a key run on one of his typical heads-up plays during a game-winning, Series-ending rally. He had maneuvered into a rundown between third and home to allow two other base runners to get into scoring position and, seeing no one covering the plate, wheeled past the catcher as he threw to third baseman Heinie Zimmerman - who unsuccessfully chased the fleet Collins home.

In 1918 Collins joined the Marines but was back the next season on another pennant-winner, the infamous 1919 Chicago "Black Sox." As one of the "honest players," he was unforgiving of the eight who had sold out, yet described the team as the greatest on which he had played, winning despite hostility, feuds, and outright crookedness.

Collins continued to play season after season of superlative second base, always batting over .300, and after the White Sox finished last in 1924 he was named manager. Collins led them for two seasons, winning more than he lost, but finished fifth both years. The White Sox judged that his days as an everyday infielder were ending, and released the $40,000-a-year player-manager.

Connie Mack then invited his former star to return to his rebuilt Philadelphia A's. Collins played less and less, but took over more and more field duties from Mack. He was third-base coach and unofficial assistant manager, and turned down offers to manage other teams.

Collins' opportunity to run a team came with the Boston Red Sox. He and Tom Yawkey were alumni of the same prep school and became friends. The millionaire sportsman, on Collins's advice, purchased the Red Sox and brought Collins in as part-owner and GM. Collins began rebuilding a team that had never recovered from the sale of stars to the Yankees a decade earlier, and although he went on just one scouting trip for the Red Sox to California he came back with two extraordinary prospects - Bobby Doerr, and a future Marine fighter pilot named Ted Williams.

Of his time in the Marines Collins wrote, "I was a buck private when I entered... I suffered every drudge any enlisted Marine suffered, but I learned from that bunch the meaning of spirit and determination... mostly it was drill and exercise and guard duty... or target practice with a snarling sergeant standing behind you... but it was an experience of which I am proud. Outside of my accomplishments on the baseball field, I am most proud of my one-time membership in the greatest of all military units - the United States Marines."

When the war ended on November 11, 1918 there was some question about Collins' availability for the 1919 season since he was still a Marine. That was resolved on February 6, when he was granted an honorable discharge from active duty - although Collins did serve four years in the Marine Corps Reserve until August of 1922. Even so, as the 1919 pennant race went into the stretch run, one group the White Sox could count on for support was the Marine Corps. A game against the Yankees was even declared "Marine Day" by owner Charles Comiskey, with Captain L. W. Putnam endorsing the event on behalf of the Corps by saying, "Eddie Collins is a Marine, and the Marines as a Corps are White Sox rooters." Two thousand Marines attended, and prior to the game they carried Collins around the grandstand on their shoulders.

ART DONOVAN
The Bulldog

Arthur Donovan, Jr. is a former football defensive tackle who was born on June 5, 1925 in the Bronx, New York and was the son of famed boxing referee Art Donovan, who supervised many of Joe Louis's bouts.

Donovan served as a Marine aircraft gunner aboard *USS San Jacinto* during WWII and saw action in the Pacific Theater from 1943 to 1945. After the war he completed his schooling at Boston College and joined the NFL in 1950 as a defensive tackle. Donovan was selected to five Pro Bowls (1953 - 1957) and also played on the world championship Baltimore teams of 1958 and 1959 - including the 1958 championship game against the New York Giants, the legendary "Greatest Game Ever Played," which became symbolic of the inception of the modern NFL. His Colts #70 jersey was retired by the team in 1966, and he was elected to the Hall of Fame in 1968.

Donovan was noted as a jovial and humorous person during his playing career, and he capitalized on that with television and speaking appearances after retiring as a player. He refers to himself as "semiretired," and notes the irony of his celebrity status. "Everywhere I go people say, 'Art, I had a beer with you.' If that were true, I would have drank every night of my life!"

Art Donovan still has a row of ribbons from World War II at home in Towson and kept the old uniform, too - though it's currently on loan to the Pro Football Hall of Fame. He hasn't worn it in decades. "I can't get my arms in the sleeves," he said.

Donovan was inducted into the Marine Corps Sports Hall of Fame in 2004 along with six-time NBA All-Star Richie Guerin and two-time heavyweight boxing champion Ken Norton. Sixty years ago, as an anti-aircraft gunner aboard *San Jacinto*, Art wondered if he'd see another sunrise, and at eighty he could still hear the drone of Japanese warplanes. "Those damn planes were like flies, comin' from all over," he said. "And the kamikazes! You almost couldn't knock them down. You'd hit 'em, they'd catch on fire... and they'd still come right at you. Some blew up so close, their parts landed on the flight deck."

Also aboard *San Jacinto* was Navy Lieutenant George H.W. Bush, a pilot whose bomber was shot down in September of 1944. Rescued at sea, Bush would go on to become the nation's 41st president.

That October *San Jacinto* was supporting the assault on Leyte in the central Philippines. When the American fleet took on the approaching Japanese fleet the light aircraft carrier became a part of the largest fleet battle in naval history, according to the U.S. Navy. Then, after spending thirteen months at sea, Donovan volunteered for the Fleet

Marine Force - which landed him in the midst of combat on Okinawa.

"I spent two months in Okinawa, and though the island was allegedly secured there were [Japanese] all over the place. You'd never really see them, you'd just hear their bullets whistling past your ear when you'd get into a fire fight."

That and other Donovan recollections appear in *Fatso*, his 1987 autobiography of a gregarious guy from the Bronx. "Artie Donovan is a true hero with an exemplary record," said Gary Bloomfield, author of *Duty, Honor, Victory*, a history of athletes who served in World War II. "He was in some of the most brutal battles of the war. You can put him right up there with anyone."

Art Donovan is presently the owner of the Valley Country Club in Baltimore, where he lives with his wife, Dorothy.

BILL FITCH
Hoops DI

William Fitch is a former NBA coach who was born on May 19, 1934 in Davenport, Iowa and has been successful in making teams playoff contenders throughout his coaching career. Before entering the professional ranks he coached college ball at the University of Minnesota, Bowling Green State University, the University of North Dakota, and his alma mater, Coe College. Fitch's teams twice qualified for the NCAA tournament.

Fitch was a former Marine Corps drill instructor, a fact that Larry Bird credited in his book *Drive: The Story of My Life* as an important reason for Bird's own strong work ethic.

During his twenty-five-year pro coaching career Fitch repeatedly was hired in an attempt to improve failing teams, and as of 2004 ranked fifth among all NBA coaches in all-time number of victories with 944.

255

Fitch was awarded as the NBA's Coach of the Year Award twice, and guided Bird, Kevin McHale, Robert Parish and the rest of the Boston Celtics to the 1981 NBA championship, defeating the Houston Rockets four games to two in the finals. From Boston Fitch went on to coach the Rockets, where he led a team featuring Hakeem Olajuwon and Ralph Sampson to the 1986 NBA Finals where they were defeated once again by Bird's Celtics, four games to two, for the NBA championship.

Fitch also coached the Cleveland Cavaliers, New Jersey Nets and Los Angeles Clippers.

HAYDEN FRY
Tigerhawk

John Hayden Fry was a National Collegiate Athletic Association Division I-A college football coach who was born on February 28, 1929 in Eastland, Texas. Fry is a descendant of one of Texas' First Families, and his great-great-grandfather fought beside General Sam Houston in the Mexican War. From 1962 to 1998 he coached at SMU and the University of Iowa, compiled a record of 232 wins, 178 losses, and 10 ties, and has been inducted into the College Football Hall of Fame.

Fry worked multiple jobs as a child to help his family through the Great Depression. He also played sports, partly to stay out of trouble. When he played safety and quarterback for Odessa High School in the 1940s their stands were routinely filled with sellout crowds. In Fry's senior year Odessa won fourteen straight games, scoring almost four hundred points and allowing about fifty, and they did not commit a single turnover all season. The Texas state playoffs placed every school into a single bracket, and at the end of the year Hayden Fry quarterbacked Odessa to the Texas state high school championship of 1946.

Fry then played at Baylor University from 1947-1950 and they had a 26-13-2 record during his four years there. Fry started a few games as an upperclassman at Baylor but could never win the full-time starting quarterback job, and he graduated with a degree in psychology in 1951.

Fry was an American history teacher and assistant football coach at Odessa High School in 1951 before joining the Marine Corps in 1952. During his time in Odessa he met and befriended a young George H. W. Bush, who would later become the 41st President of the United States.

Hayden Fry served in the Marine Corps from 1952-1955. He played with the Quantico Marines football team in 1953, winning the Marine Corps championship and playing in the Poinsettia Bowl. He also coached a six man football team while in the Marines, and the unique style of play allowed him to innovate and invent new schemes. He became friends with Al Davis, who was at the time coaching a rival military team and would later become the famous owner of the Raiders. Hayden Fry was discharged from the Marines in February of 1955 with the rank of captain.

In 1955 Fry returned to Odessa as a teacher and assistant football coach. The following season Odessa head coach and former Texas A&M freshmen varsity coach Cooper Robbins was promoted to athletic director, and Hayden Fry took his first head coaching job. At twenty-six years of age, he was coaching the high school he had led to the state title less than ten years earlier. He served as Odessa's head coach for three years and during that time met and befriended the head coach at Texas A&M, Bear Bryant, and one of his history students, Roy Orbison, who later became a musical star.

After stints as an assistant at Baylor and Arkansas Southern Methodist University tapped Fry as their next head football coach for the 1962 season when he was just thirty-one years old. The SMU Mustangs were members of the Southwest Conference at the time, and Fry won the conference coach of the year award in his first season.

When Fry took the job at SMU he was promised that he would be allowed to recruit black athletes. Fry and the school wanted to make certain that the first player they recruited was not only a good athlete, but also a good student and citizen and someone with the mental toughness to be the first black player in conference history. Fry found that player in Jerry LeVias. He was a great player, an exceptional student, and mentally tough. He had never had discipline problems, and was deeply religious - the perfect player for SMU. Fry received abuse for recruiting a black player to SMU in the form of hate mail and threatening phone calls, but he downplayed the treatment because the harassment of LeVias was much, much worse.

Hayden Fry compiled a 49-66-1 record in eleven seasons at SMU, including the school's only three winning seasons since the late 1940s. In his autobiography he stated that he believed his firing after the 1972 season was related to several boosters' desire to start a slush fund to pay players and recruits, because since SMU was the second-smallest school in the Conference it was difficult to compete against schools double its or more size. When he refused to go along with the plan the boosters pressured the school's new president to fire him, and as it turned out SMU would be hit with NCAA sanctions five times after Fry's departure before having its program completely shut down for the 1987 season due to a massive litany of misconduct.

After six successful seasons in North Texas Hayden Fry was hired as Iowa's head football coach after the 1978 season. He had never been to Iowa, but knew and liked the athletic director there. Iowa had had seventeen straight non-winning seasons, but Fry was impressed by the fan support for a program that had struggled for so long.

Fry turned his attention to changing a losing attitude and starting new traditions at Iowa. He would not celebrate close losses or moral victories, and hired a marketing group to create the "Tigerhawk" logo to represent the University of Iowa. Since both shared the colors of black and gold, Fry gained permission from the Pittsburgh Steelers, the dominant NFL program of the time, to overhaul Iowa's uniforms in the Steelers' image. Fry had the team "swarm" onto the field together as they left the locker room, holding hands in a show of solidarity. He also had the visitors' locker room painted pink. Fry, a psychology major at Baylor, knew that pink is occasionally used in jails and mental institutions to relax and pacify the residents and claimed it might have the same effect on visiting teams. Principally, though, Fry hoped the unusual color would distract and fluster opposing players and coaches.

Iowa had losing seasons in 1979 and 1980, and some began to wonder if Fry would suffer the same fate as the four coaches before him who had left Iowa after failing to produce a winning season - but the team broke through in 1981 in a magical season for Hawkeye fans. Iowa began the year by upsetting sixth ranked Nebraska, a team that had defeated them 57-0 the previous season. Two weeks later they defeated sixth ranked UCLA, and later that season the Hawkeyes beat Michigan in Ann Arbor for their first victory over the Wolverines in nineteen years. Then a victory over Purdue snapped a twenty game losing streak to the

Boilermakers and clinched Iowa's first winning season in nineteen years.

In the final game of the 1981 regular season Iowa's win over Michigan State, coupled with an Ohio State upset of Michigan in Ann Arbor, gave Iowa a share of the 1981 Big Ten title. Either Michigan or Ohio State had gone to the Rose Bowl in each of the previous twelve seasons, prompting critics to nickname the Big Ten the "Big Two and Little Eight." Fry had altered the balance in the league by leading the Hawkeyes to a share of the league title and a berth in the 1982 Rose Bowl. More success would follow.

In 1989 the television show *Coach* debuted starring Craig T. Nelson as fictional football coach "Hayden Fox." The title character, created by Iowa alumnus Barry Kemp, was loosely based on Fry, and he later appeared in commercials for the NCAA with the female lead of the series, Shelley Fabares.

The 1998 season marked Fry's twentieth at Iowa, and it was also his worst as the Hawks finished with a 3-8 record and a home loss to intrastate rival Iowa State (Fry's first loss to them in fifteen years). It would also be his *last* season at Iowa as Fry, who was secretly undergoing radiation treatments for prostate cancer all year, announced his retirement on November 22, 1998.

It is difficult to overstate Hayden Fry's positive impact on Iowa football. He had a 143-89-6 record at Iowa and led the Hawkeyes to three Big Ten titles, three Rose Bowl appearances, and fourteen bowl games - but more than that, he established a winning tradition both on and off the field.

After successful prostate cancer treatment Hayden Fry retired to Nevada and was inducted into the College Football Hall of Fame in 2003 alongside former SMU star Jerry LeVias.

ERNIE HARWELL
Voice of the Tigers

William Earnest "Ernie" Harwell is a former Major League Baseball sportscaster who was born on January 25, 1918 in Washington, Georgia. For fifty-five years, forty-two of them with the Detroit Tigers, Harwell called balls, strikes, and home runs on radio and television, and in January of 2009 the American Sportscasters Association ranked him sixteenth on its list of the Top Fifty Sportscasters of All Time. In his fifty-five-year career he missed only two games – one for his brother's funeral in 1968, and the other to attend his induction into the National Sportscasters and Sportswriters Association Hall of Fame in 1989.

Harwell grew up in Atlanta, working in his youth as a paperboy for the *Atlanta Georgian*, and one of his customers was *Gone With the Wind* author Margaret Mitchell. He was an avid baseball fan from an early age, became visiting batboy for the Atlanta Crackers of the Southern Association at the tender age of five, and has never had to buy a ticket for a baseball game since then.

After graduating from Emory University (where he helped edit *The Emory Wheel*), Harwell began his career as a copy editor and sportswriter for the *Atlanta Constitution* and as a regional correspondent for *The Sporting News*. Then he began announcing games for the Crackers on WSB radio, after which he served four years in the Marine Corps from 1942 to 1946.

In 1948, Harwell became the only announcer in baseball history to be traded for a player when Brooklyn Dodgers' general manager Branch Rickey traded catcher Cliff Dapper to the Crackers in exchange for Harwell's broadcasting contract. (Harwell was brought to Brooklyn to substitute for regular Dodger announcer Red Barber, who was hospitalized with a bleeding ulcer.)

He broadcast for the Dodgers through 1949, the New York Giants from 1950-1953 (including his call of Bobby Thomson's "shot heard 'round the world" in the 1951 National League pennant playoff game on NBC television), and the Baltimore Orioles from 1954–1959.

In 1960 Harwell became the "voice" of the Tigers, replacing Van Patrick. Among his partners, for a period of nineteen seasons, was Paul Carey. He then worked a part-time schedule for the California Angels in 1992, and the following year returned toTigers' radio, calling innings 1–3 and 7–9 of each game. From 1994 to 1998 Harwell called television broadcasts for the Tigers, and in 1999 resumed full-time radio duties with the team and continued in that role through 2002. During spring training of that year Harwell announced he would retire at the end of the season, with his final broadcast coming on September 29, 2002.

Nationally, Harwell broadcast two All-Star Games and two World Series for NBC Radio, numerous ALCS and ALDS series for CBS Radio and ESPN Radio, and the CBS Radio *Game of the Week* from 1992 to 1997. He also called the 1984 World Series for the Tigers.

Following his retirement, Harwell came back briefly in 2003 to call a *Wednesday Night Baseball* telecast on ESPN as part of that network's "Living Legends" series of guest announcers. In 2005 he also sat in for an inning on the FOX network's coverage of the All-Star Game (which was held in Detroit that year), as well as an inning on the ESPN Radio broadcast.

Harwell also appeared as a guest on an ESPN *Sunday Night Baseball* telecast in Detroit on July 1, 2007, and his typical sense of humor was on display. He talked about working beside the deep-voiced Paul Carey ("next to him, everyone sounds like a soprano") for nineteen years, "which seemed like thirty." He then asked Jon Miller and Joe Morgan how long they had worked together. "Nineteen years" was the response. Harwell grinned at both of them, "Uh-huh, uh-huh."

Harwell currently does occasional vignettes (small video clips) on the history of baseball for Fox Sports Detroit's magazine program *Tigers Weekly*. A devout Christian, he has also long been involved with the Baseball Chapel, an evangelistic organization for professional ballplayers.

In 2004 the Detroit Public Library dedicated a room to Ernie Harwell and his wife which will house his collection of baseball memorabilia which is valued at over two million dollars. He has also been inducted into the Michigan Sports Hall of Fame, the Georgia Sports Hall of Fame, and the National Sportswriters and Sportscasters Association Hall of Fame, and in 1981 became the fifth man to be honored with the Ford C. Frick Award by the Baseball Hall of Fame. He has also received the Ty Tyson Award for Excellence in Sports Broadcasting from the Detroit Sports Broadcasters Association. Harwell has written seven books about baseball, and produced an audio scrapbook about his career in 2006. He has also written sixty-six songs, once quipping that he had "more no-hitters than Nolan Ryan."

Ernie Harwell ended his final broadcast on September 29, 2002, with the words, "I thank you very much, and God bless all of you." He currently lives in Novi, Michigan with his wife Lulu. Now in his nineties, Harwell was diagnosed with an incurable cancer of the bile duct on September 3, 2009, and gave a sentimental farewell speech before a game at Comerica Park two weeks later.

ELROY HIRSCH
Crazy Legs

Elroy "Crazylegs" Hirsch (June 17, 1923 - January 28, 2004) was a football running back and receiver who was born in Wausau, Wisconsin. He played for the Los Angeles Rams and Chicago Rockets, and was nicknamed for the usual running style he developed by running cross legged over four square cement sidewalk blocks in his hometown.

Hirsch played his first collegiate season with the University of Wisconsin Badgers in 1942. His nickname was permanently affixed to him by *Chicago Daily News* sportswriter Francis Powers who, upon witnessing him play for the Badgers against the Great Lakes Naval Station in 1942 wrote, "His crazy legs were gyrating in six different directions, all at the same time. He looked like a demented duck."

His commitment to the Navy V-12 program of the United States Marine Corps required him to transfer to the

University of Michigan. Hirsch played two intercollegiate football seasons for the Michigan Wolverines, where during the 1943-44 year he earned the distinction of being the only athlete at the school to letter in four sports (football, basketball, track and baseball) in a single year. He was inducted into the College Football Hall of Fame in 1974.

Hirsch was drafted by the Chicago Rockets of the All-America Football Conference, where he played from 1946 to 1948 in three injury-prone seasons. After the Rockets and the AAFC merged with the NFL he joined the Los Angeles Rams through 1957, where he gained his fame. Coach Clark Shaughnessy made Hirsch the first full-time 'flanker' in NFL history, splitting the talented receiver outside from his previous halfback position. Additionally, he was one of the first to sport the molded plastic helmet that is the industry standard today in the NFL. Coach Shaughnessy had it fitted him for him as a precaution, as he was injured when first joining the Rams. When Hirsch was playing for Chicago in an All-America game against the Cleveland Browns he got tackled so badly that his right knee ligaments were torn and he suffered a fractured skull above his right ear as well.

Hirsch was a key to the Rams victory in the 1951 NFL championship with an NFL record 1,495 yards receiving which stood for nineteen years. He also had sixty-six catches and seventeen touchdowns that same year, and was inducted into the Pro Football Hall of Fame in 1968 with a career 387 receptions, 7,029 yards, and sixty touchdowns.

Elroy Hirsch served as the Director of Athletics for the University of Wisconsin-Madison from 1969 to 1987, and died of natural causes at an assisted living home in Madison, Wisconsin on January 28, 2004.

GIL HODGES
The Miracle Worker

Gilbert Raymond "Gil" Hodges (April 4, 1924 - April 2, 1972) was a Major League Baseball first baseman and manager who was born in Princeton, Indiana, the son of coal miner Charlie and his wife Irene. He played most of his career for the Brooklyn and Los Angeles Dodgers and was the major leagues' outstanding first baseman of the 1950s, with teammate Duke Snider being the only player to have more home runs or runs batted in during that decade. His 370 career home runs set a National League record for right-handed hitters and briefly ranked tenth in major league history, and he also held the NL record for career grand slams from 1957 to 1974. Hodges anchored the infield on six pennant winners, and remains one of the most beloved and admired players in team history. A sterling defensive player, he won the first three Gold Glove Awards and led the NL in double plays four times and in putouts, assists and fielding percentage three times each. He ranked second in NL history with 1,281 assists and 1,614 double plays when his career ended, and was among the league's career leaders in games and total chances at first base. He also managed the New York "Miracle" Mets to the 1969 World Series title in one of the greatest upsets in Series history before his untimely death in 1972.

Hodges was a star four-sport athlete at Petersburg High School, earning a combined seven varsity letters in football, baseball, basketball, and track. He declined a 1941 contract offer from the Detroit Tigers, and instead attended Saint Joseph's College with the hope of eventually becoming a collegiate coach. He was signed by the Brooklyn Dodgers in 1943, and appeared in one game for the team as a third baseman that year.

Hodges entered the Marine Corps during World War II after having participated in its ROTC program at Saint Joseph's, served as an anti-aircraft gunner in the battles of Tinian and Okinawa, and received a Bronze Star and a commendation for courage under fire for his actions. After his 1946 discharge he returned to Brooklyn and saw time as a catcher in 1947, joining the team's already solid nucleus of Jackie Robinson, Pee Wee Reese and Carl Furillo - but the emergence of Roy Campanella made it evident Hodges had little future behind the plate so he was shifted by manager Leo Durocher to first base where his play came to be regarded as exemplary.

Hodges was an eight-time All-Star, from 1949-55, and again in 1957. With his last home run of 1952 he tied Dolph Camilli's Dodger career record of 139, and he passed him in 1953 - although the great Duke Snider would move ahead of him in 1956. He again led the NL with 116 assists in 1952, and was third in the league in home runs and fourth in RBI and slugging. A great fan favorite in Brooklyn, he was perhaps the only Dodger regular who was never booed at their home park of Ebbets Field. The fans were very supportive even when Hodges suffered through one of the most famous slumps in baseball history, going hitless in the last nine games of 1952. Then, during the 1952 World Series against the Yankees, he finished the Series 0-21 at the plate as Brooklyn lost in seven games. When his slump continued into the following spring fans reacted with countless letters and good-luck gifts, and one Brooklyn priest - Father Herbert Redmond of St. Francis Roman Catholic Church - even told his flock, "It's far too hot for a homily. Keep the Commandments, and say a prayer for Gil Hodges." Hodges began hitting again soon afterward, and rarely struggled again in the World Series.

After his playing career was over Hodges became a manager. He guided the Washington Senators through 1967, and although they improved in each season the team never achieved a winning record. One of the most notable incidents in his career occurred in the summer of 1965 when pitcher Ryne Duren, who was nearing the end of his career and sinking into alcoholism, walked onto a bridge with the intention of committing suicide. Hodges talked him away from the edge.

In 1968 Hodges was brought back to manage the perennially woeful Mets, and while the team only posted a 73-89 record it was nonetheless the best mark in their seven-year existence. Then in 1969 he led the "Miracle Mets" - anchored by star pitcher and fellow Marine Tom Seaver - to a World Series championship by defeating the heavily favored Baltimore Orioles. After finishing higher than ninth place for the first time ever, the Mets became not only the first expansion team to win the Series, but also the first team to win the Series after finishing at least fifteen games under .500 the previous year, and Hodges was named *The Sporting News'* Manager of the Year.

The moment many Mets fans of a certain age and Hall of Fame slugger and Mets' announcer Ralph Kiner consider the most memorable in team history - and the turning point in the team's 1969 season - came in the third inning of the second game of a July 30th doubleheader against the Houston Astros. When Mets' star left fielder Cleon Jones failed to hustle after a ball hit to the outfield Hodges removed him from the game - but rather than simply signal from the dugout for Jones to come out, or delegate the job to one of his coaches, Hodges left the dugout and walked slowly and deliberately all the way out to left field to remove Jones and then walked him back to the bench. For the rest of that season Jones never again failed to hustle. Kiner retold that story dozens of times during Mets broadcasts, both as a tribute to Hodges, and as an illustration of his quiet but disciplined character.

After identical third-place seasons of 83-79 in 1970 and 1971 Hodges died suddenly of a heart attack on April 2, 1972 in West Palm Beach, Florida while playing golf with other members of the Mets coaching staff, including Yogi Berra, during an off day from spring training. He was survived by his wife, the former Joan Lombardi, whom he had married on December 26, 1948, and his son and three daughters.

Gil Hodges is buried at Holy Cross Cemetery in East Flatbush, Brooklyn. The Mets wore a black-armband on the left sleeves of their uniform jerseys during the 1972 season in his honor, and his in one of only three numbers ever to be retired by the team.

KEITH JACKSON
Kickoff Keith

Keith Jackson is a former sportscaster who was born on October 18, 1928 on a farm outside of Carrollton, Georgia near the Georgia-Alabama state line where he grew up. He was best known for his long career with ABC Sports television and his coverage of college football, as well as his style of folksy, down-to-earth commentary and deep voice.

The only surviving child in a poor family, he grew up listening to sports on the radio. After a stint in the Marine Corps he attended Washington State University under the G.I. Bill. Jackson began as a political science major, but while there became interested in broadcasting and graduated in 1954 with a degree in Speech Communications.

Though best known for his college football broadcasts, Jackson announced numerous other sports for ABC throughout his career including Major League Baseball,

NBA basketball, boxing, auto racing, the USFL, and the Olympic Games. He also briefly worked college basketball with Dick Vitale, and served as the anchor for ABC's coverage of Super Bowl XXII in 1988.

Jackson began his career as a broadcaster in 1952 when he called a radio game between Stanford and Washington State. He also covered hydroplane races, minor league Seattle Rainiers baseball games, and University of Washington Huskies football games. In 1958 Jackson became the first American sports announcer to broadcast an event from the Soviet Union, a crew race between the Washington Huskies and a Soviet team. Despite heavy suspicion and numerous hurdles by the Soviet authorities Jackson and his cohorts were able to cover the race and witness the first ever American sports victory on Russian soil. He joined ABC Sports in 1962, went full time in 1964 as a radio news correspondent, and later became sports director of ABC Radio West.

In the early 1960s Jackson covered American Football League games. Then in 1970 he was chosen to be the first play-by-play announcer for *Monday Night Football*, but only remained in that capacity for the program's first season. Frank Gifford had been ABC's initial target, but he could not get out of his CBS contract until after the 1970 season. In 1971 Gifford landed the job, and Jackson found out he had been taken off the *Monday Night* package from thirty-eight messages rather than from Roone Arledge himself. This led to some contention between him and the brass at ABC.

Jackson was involved in the ABC coverage of the 1972 Summer Olympics, and continued to contribute even when an attack by Palestinian terrorists transformed the broadcast from that of a typical sporting event to that of a greater international and historical news event. In all he covered a total of ten Summer and Winter Olympic Games.

For all his success, Jackson received the most acclaim for his coverage of college football. He genuinely enjoyed the sport, liked the purity of it, and began broadcasting at a time when television play-by-play announcers did not always have regular analysts. He would only miss working a college season once in his over fifty years from 1952 on, and that was because he was serving as play-by-play announcer during the inaugural season of *Monday Night Football.*

Jackson announced his retirement from college football at the end of the 1998 season, with the intention of living full time at his home in California. Choosing the 1999 National Championship at the Fiesta Bowl between Tennessee and Florida State as his last broadcast, he concluded the program by stating, "Tennessee 23, Florida State 16. And so it is done. I say goodbye to all of you. God bless and good night."

He rescinded this decision the following fall and began to do a more limited schedule of games teamed with Tim Brant and later Dan Fouts, almost exclusively sticking to venues on the West Coast which were closer to his home in British Columbia. One of the notable exceptions was the 2003 Michigan-Ohio State game, the one hundredth meeting between the two archrivals. He strongly hinted that he was interested in retiring for good after the 2005 season, telling *The New York Times* that he was feeling his age after fifty-three seasons. ABC tried to convince Jackson to stay, but he had made up his mind.

Keith Jackson retired for good on April 27, 2006 at age seventy-seven, noting that he "didn't want to die in a stadium parking lot." He and his wife Turi Ann have three grown children and homes in the Los Angeles, California area and Pender Harbour, British Columbia in Canada.

MILLS LANE
Maximum Mills

Mills Bee Lane III is a boxing referee, boxer, judge and television personality who was born on November 12, 1936 in Combalee, South Carolina. He is best known for having officiated several major heavyweight championship boxing matches in the 1980s and 1990s and for starring in the television show *Judge Mills Lane*.

Lane hails from a prominent Georgia family. His grandfather founded the largest bank in Georgia, and his uncle (and namesake) was the president of Citizens & Southern National Bank. Lane, however, had other aspirations, and joined the Marine Corps in 1956 after his graduation from Middlesex School. He started boxing while in the Marines, and became the All-Far East welterweight champ. After leaving the Corps he enrolled at the University of Nevada, Reno and became the NCAA boxing champion, amassing an amateur record of 60-4. He turned pro while in

275

college, eventually posting an 11-1 record , and was in the 1960 Summer Olympics boxing finals in San Francisco where he was defeated by Phil Baldwin in the semifinals.

Lane graduated with a business degree in 1963, and then a few years later enrolled at the University of Utah to attend law school. Lane graduated as a lawyer, and later on became a prosecutor at the Washoe County district attorney's office in Reno. In 1979 he became Chief Deputy Sheriff of Investigative Services at the Washoe County Sheriff's Office, and was later elected District Attorney in 1982 and District Judge in 1990.

Lane refereed his first world championship boxing match in 1971 when Betulio Gonzalez had a fifteen-round draw with Erbito Salavarria for the WBC flyweight title. Lane then became a household name in the United States the night he refereed "The Bite Fight" rematch between world heavyweight champion Evander Holyfield and challenger Mike Tyson on June 28, 1997. In that infamous bout Lane disqualified Tyson after he bit Holyfield's ears twice. Lane's shirt was stained with blood from the incident, and he sold it to a memorabilia collector on the very same night. The following year after the fight between Thomas Hearns and Jay Snyder on November 6, 1998, Mills Lane retired from being a boxing referee.

From 1998 to 2001 his court show *Judge Mills Lane* aired on national television. In addition to this show the producers of MTV's *Celebrity Deathmatch* approached him about having his character and voice used in their show as the referee of their plasticine figure matches. Lane accepted the offer, and thereby became an MTV personality. As a referee Lane had started boxing matches by declaring, "Let's get it on!," which became his catchphrase. This phrase was reproduced in *Celebrity Deathmatch*. Lane later named his autobiography *Let's Get It On: Tough Talk from Boxing's Top Ref and Nevada's Most Outspoken Judge.*

Mills Lane suffered a debilitating stroke in March of 2002 which left him partially paralyzed. His adopted city of Reno celebrated him on December 27, 2004 by proclaiming it "Mills Lane Day," and in January of 2009 he made his first public appearance in years at the dedication of a new courthouse which is named after him.

TOMMY LOUGHRAN
The Fighting Marine

Tommy Loughran (November 29, 1902 - July 7, 1982) was the light heavyweight boxing champion of the world. His effective use of coordinated footwork, sound defense, and swift, accurate counter punching is now regarded as a precursor to the techniques practiced in modern boxing. Loughran fought many middleweight, light heavyweight, and heavyweight champions in his career - including Gene Tunney, Jack Sharkey and Georges Carpentier. He even achieved a Newspaper Decision over fistic phenom Harry Greb despite being only nineteen years old when they first met. As a light heavyweight he defeated two future world heavyweight champions, Max Baer and James J. Braddock.

When Loughran was in his prime in the late 1920s it was said he could fight fifteen rounds without disturbing a single wavy hair on his handsome head. Born Thomas Patrick Loughran on November 29, 1902 in Philadelphia, he was the son of an Irish immigrant who married a Philadelphia-born woman. He was the quintessential Irish fighter, and the pride of South Philadelphia. His deportment in and out of the ring was pure class. Even decades later old-timers remembered him as the most stylish of fighters, a beautiful boxer, and the most gracious of men.

Loughran's boxing career spanned nineteen years from 1919 to 1937, and though the Ring Record Book lists 172 fights under his name he probably fought a lot more than that. He won the light-heavyweight championship in 1927, and successfully defended the title six times before moving up to the heavyweight division in 1929. He fought once for the heavyweight championship against giant Primo Camera in 1934 when the signs to retire were beginning to appear. He was to say later, "When one minute between rounds isn't enough time to recuperate, then you know." Although he lost on a decision, the fight went into the *Guinness Book of World Records* for having the greatest weight disparity of any heavyweight championship bout - Camera's 270 pounds to Loughran's 184.

Loughran was an intelligent fighter and was very articulate, even as a schoolboy. He graduated from high school three years early, primarily because of his exceptional memory. He was also ahead of his age group in physical maturity and joined the Marines at fourteen, telling the recruiter he was twenty-three. In the Marines, they put boxing gloves on him for the first time. "They matched me with a tough Marine, and what I did to him was nobody's business," he told an interviewer many years later.

Tommy Loughran died on July 7, 1982 in Altoona, Pennsylvania. He is a member of the International Boxing Hall of Fame, and was the *Ring Magazine* Fighter of the Year in 1929 and 1931.

BOB MATHIAS
America's Goodwill Ambassador

Robert Bruce Mathias (November 17, 1930 - September 2, 2006) was a decathlete, two-time Olympic gold medalist, and United States Congressman who was born in Tulare, California.

Mathias first took up the decathlon at the suggestion of his coach, Virgil Jackson, at Tulare High School early in 1948, and during the summer qualified for the United States Olympic team for the 1948 Summer Olympics in London.

In the Olympics Mathias' naïveté for the decathlon was exposed. He was unaware of the rules in the shot put, and nearly fouled out of the event. He almost failed in the high jump, but was able to recover. Even so Mathias overcame his difficulties and won the Olympic gold medal easily, and at the age of seventeen became the youngest gold medalist to win a track and field event.

280

Mathias continued to fare well in decathlons during the four years between the London games and the 1952 Summer Olympics in Helsinki. In 1948 he won the James E. Sullivan Award as the nation's top amateur athlete, but because his scholastic record in high school did not match his athletic achievements he spent a year at the Kiski School, a well respected all-boys boarding school in Saltsburg, Pennsylvania. Mathias then entered Stanford University in 1949, played college football for two years, was a member of Phi Gamma Delta fraternity, set his first decathlon world record in 1950, and led Stanford to a Rose Bowl appearance in 1952. Also during that period, in 1951, Mathias spent the summer going through Marine Corps boot camp in San Diego.

At Helsinki Mathias asserted himself as one of the world's best athletes. He won the decathlon by 912 points, an astounding margin, became the first to successfully defend an Olympic decathlon title, and returned to the United States as a national hero. In 1952 he was, therefore, the first person to ever compete in an Olympics and a Rose Bowl in the same year.

After the 1952 Olympics Mathias retired from athletic competition and later became the first director of the United States Olympic Training Center - a post he held from 1977 to 1983.

In 1954 Bob Mathias entered active duty in the Marine Corps as a second lieutenant, and from 1954 to 1956 he visited more than forty countries as America's Goodwill Ambassador. Also in 1954 a film about his early life called *The Bob Mathias Story* was released in which he and his wife Melba played themselves. He also starred in a number of mostly cameo-type roles in a variety of movies and TV shows throughout the 1950s, and during 1959-1960 television season played Frank Dugan, with costars Keenan Wynn as Kodiak and Chet Allen as Slats, in the NBC adventure series *The Troubleshooters*.

Between 1967 and 1975 Matthias served four terms in the United States House of Representatives as a Republican Congressman representing the northern San Joaquin Valley of California. He initially defeated fourteen-year incumbent Democrat Harlan Hagen by eleven points in 1966, which was not surprising since the area had started pulling away from its New Deal Democratic roots.

Mathias was reelected three times without serious difficulty, but in 1974 his district was radically altered in a mid-decade redistricting. Most of the more rural parts of his district were cut out and replaced with a portion of Fresno and he was narrowly defeated by Fresno County Supervisor John Hans Krebs and became one of several Republicans to be swept out in the wake of the Watergate scandal.

Bob Mathias died of cancer in Fresno, California on September 2, 2006 at the age of seventy-five.

"TUG" MCGRAW
Ya Gotta Believe!

Frank Edwin "Tug" McGraw Jr. (August 30, 1944 - January 5, 2004) was a Major League Baseball relief pitcher who was born in Martinez, California. The father of country music singer Tim McGraw and actor/TV personality Mark McGraw, he gained stardom during the New York Mets victory in the 1969 World Series and is likely best remembered for coining the motto "Ya Gotta Believe" during the Mets' run for the 1973 World Series. He is also renowned as the star reliever who pitched the final strike for the 1980 World Champion Philadelphia Phillies.

Tug graduated from St. Vincents High School in Vallejo, California in 1962 and enrolled in a local barber college where his poor scissor technique earned him the nickname "Tug." The Mets signed him as an amateur free agent on June 12, 1964 and initially tried him as a starting pitcher, but he only managed a 2-12 record in sixteen starts over two years - although one of those victories was against the legendary Sandy Koufax and marked the first time the Mets had ever beaten the future Hall of Famer. Relying on a good screwball, he racked up twelve saves for the 'Miracle Mets' as they went on to win the World Series, although he did not pitch in the Fall Classic.

PFC McGraw reported to MCRD Parris Island on September 23, 1965 along with fellow New York Met pitcher Jim Bethke and was ultimately trained as a reserve rifleman and marksman skilled in the use of the M-14 rifle and M-60 machine gun. McGraw would get his infantry training at Camp Lejeune where he, in his own words, became a "trained killer." Many of the skills which he acquired during his training translated very positively towards his career with the New York Mets and later the Philadelphia Phillies, since the Marines taught McGraw discipline, concentration, and confidence - all things which translated seamlessly onto the pitcher's mound.

For McGraw, one of the most challenging aspects of being in the military was the internal conflict which it stirred within him. At the same time he was finishing his Marine training his brother, Dennis McGraw, was staging anti-war protests at Vallejo Junior College in California, which Tug also attended. In a March 5, 1967 *New York Times* article McGraw even admitted that he and his brother would have arguments over the way the Vietnam War was being conducted.

McGraw went on to help the Mets win the 1969 World Series, but it was in the build-up to the 1973 post-season that his signature phrase, "Ya gotta believe!" was coined. In August of that year the Mets were down more than eleven games and, after a particularly bad performance, Mets' Chairman M. Donald Grant delivered a torrid locker-room lecture to the chastened team. McGraw was said to have uttered the phrase to poke fun at Grant's pep talk as he emerged from the meeting, but his teammates burst out laughing and went on to a winning streak that landed them in the World Series. Even though the Mets lost to Oakland in seven games, "Ya gotta believe!" became the catchphrase of the season and would remain indelibly associated with McGraw's high-spirited personality.

Besides his pitching talents, Tug McGraw was a colorful character off the field. He once famously said, "Ninety percent (of my salary) I'll spend on good times, women, and Irish whiskey. The other ten percent I'll probably waste." Known for his spontaneous quips and graciousness to his fans, he became one of the sport's most beloved characters. "He wore his sandy hair long," noted *New York Times* writer Frank Litsky, "and with his little-boy face and boyish enthusiasm he was a crowd favorite. After a third out he would run off the mound, slapping his glove against a thigh. After a close call, he would pat his heart."

Traded to Philadelphia in 1974, he went on to help the Philles take East Division titles in 1976, 1977 and 1978 and the National League pennant in 1980 and 1983 - but it was Game Six of the 1980 World Series in that would define McGraw's career and make him a hero in his adopted home-town. In the ninth inning he struck out Kansas City's Willie Wilson with bases loaded, and the Phillies won the World Series for the first time in their ninety-seven year history.

The photograph taken just after that moment showed McGraw jumping off the mound, hands high in the air, and became one of the classic images in sports history. Another timeless photo was captured just seconds later when Phillies third-baseman Mike Schmidt jumped into his arms on the mound. Schmidt later said the two had planned it on their ride to Veterans Stadium that night. "Both of us knew whoever was on or near that mound for the final out would probably be on the cover of *Sports Illustrated*," Schmidt told the same publication. "Sure enough, it worked. Tug struck out Wilson and then turned to look at me at third base. Of course I came running and jumped on him."

The 1984 season was McGraw's last in baseball. He retired with a 96-92 record and a 3.14 earned-run average, became a television reporter for a Philadelphia station, wrote three children's books, and remained a fan favorite. The father of two sons and a daughter, Tug belatedly discovered his fourth and oldest child after an eleven-year-old Louisiana boy came across his birth certificate. Tim Smith was an ardent baseball fan, and was stunned to find the name of one of his heroes in the space on the document that listed the father's name. Smith, who later took his father's name, was the product of a summer romance between McGraw and Betty Trimble that occurred during his minor-league career - and McGraw had never known of the boy's existence. McGraw and his long-lost son enjoyed a close relationship, and Tim McGraw grew up to become a country-music legend and husband of fellow Nashville star Faith Hill.

On March 12, 2003 McGraw was working as a spring training instructor for the Phillies when he was hospitalized with a brain tumor. When surgery was performed to remove it, the tumor was revealed to be malignant and inoperable. Given three weeks to live by doctors, Tug managed to survive nine months. During that time he attended the closing ceremonies of Veterans Stadium, where he recreated the final out of the Phillies' World Series triumph.

Tug McGraw lost his battle with cancer on January 5, 2004 and was cremated after his death. Nearly five years later his son Tim took a handful of his dad's ashes and spread them on the pitcher's mound at the Phillies current home park, Citizens Bank Park, before Game Three of the 2008 World Series. The Phillies won the game, defeating the Tampa Bay Rays 5-4 en route to the team's second World Series Championship.

BILLY MILLS
A Lakota Journey

William Mervin "Billy" Mills is a Native American Olympic Gold Medalist who was born on June 30, 1938 in Pine Ridge, South Dakota and was raised on the Pine Ridge Indian Reservation. An Oglala Lakota (Sioux), he is the second Native American ever to win an Olympic gold medal. He accomplished this feat in the 10,000 meter run at the 1964 Tokyo Olympics in Tokyo, making him the only American to ever win the Olympic gold in that event, and his victory is considered one of the greatest Olympic upsets in history.

An orphan at the age of thirteen, Mills took up running while attending the Haskell Institute in Lawrence, Kansas, which is now known as Haskell Indian Nations University. Both a boxer and a runner in his youth, he eventually gave up boxing to focus on running.

Mills attended the University of Kansas on an athletic scholarship, was named an NCAA All-America cross-country runner three times, and in 1960 won the individual title in the Big Eight cross-country championships - and Kansas won the 1959 and 1960 outdoor national championship while he was on the team. After graduating with a degree in Physical Education Mills entered the Marine Corps and was a First Lieutenant in the Reserves when he competed in the 1964 Olympics.

Billy Mills qualified for the 1964 Summer Olympics on the U.S. Track and Field Team in the 10,000 meter run and marathon. The favorite in the 10K was Ron Clarke of Australia, who held the world record. The runners expected to challenge him were defending champion Pyotr Bolotnikov of the Soviet Union and Murray Halberg of New Zealand, who had won the 5000 meter run in 1960.

Mills was a virtual unknown, had finished second in the U.S. Olympic trials, and his time in the preliminaries was a full minute slower than Clarke's. Indeed, Clarke set the tone of the race, and his tactic of surging every other lap appeared to be working. Halfway through the race only four runners were still with him - Mohammed Gammoudi of Tunisia, Mamo Wolde of Ethiopia, Kokichi Tsuburaya of Japan, and Mills. Tsuburaya, the local favorite, lost contact first, and then Wolde, and with two laps to go only two runners were still with Clarke. On paper it seemed to be his race, since he had run a world record time of 28:15.6 while neither Gammoudi nor Mills had ever run under 29 minutes.

Mills and Clarke were running together with Gammoudi right behind them as they entered the final lap. They were lapping other runners and, down the backstretch, Clarke was boxed in. He pushed Mills once, and then again. Then Gammoudi pushed them both and surged into the lead as they rounded the final curve. Clarke recovered and began chasing Gammoudi, while Mills appeared to be too far back to be in contention. Clarke failed to catch the Tunisian but Mills sprinted past them both, and his winning time of 28:24.4 was almost fifty seconds faster than he had ever run before and set a new Olympic record for the event.

After the race Mills talked with Clarke and asked if he was straining as hard as he could on the final straightaway to the finish, to which Clarke simply replied, "Yes." Mills has stated that he tried to be relaxed during his final kick to the finish line and felt that helped him pass both Gammoudi and Clarke. An infrequently mentioned fact is both Clarke and Mills ran the marathon at the 1964 Olympics after the 10,000 meters, with Clarke finishing in ninth place and Mills in fourteenth in a respectable 2:22:55.4, approximately two-and-a-half minutes behind Clarke.

Mills later set U.S. records for 10,000 meters (28:17.6) and the three mile run, and had a 5,000 meter best of 13:41.4. Then in 1965 he and Gerry Lindgren both broke the world record for the six mile run when they finished in a tie at the U.S. AAU nationals while running 27:11.6.

Billy Mills was inducted into the United States Track and Field Hall of Fame in 1976 and the U.S. Olympic Hall of Fame in 1984. He is also in the National Distance Running Hall of Fame, the Kansas Hall of Fame, the South Dakota Hall of Fame, the San Diego Hall of Fame, and the National High School Hall of Fame. He is the subject of the 1984 film *Running Brave* starring Robby Benson, and serves as the spokesperson for Running Strong for American Indian Youth, an organization which helps support projects benefitting the American Indian people, especially the youth. In 1990 he wrote *Wokini: A Lakota Journey to Happiness and Self-Understanding* with Nicholas Sparks, and his book *Lessons of a Lakota* was published in 2005.

RICK MONDAY
Saving the Flag

Robert James "Rick" Monday, Jr. is a former Major League Baseball centerfielder and current broadcaster who was born on November 20, 1945 in Batesville, Arkansas. From 1966 through 1984 Monday played for the Kansas City/Oakland Athletics, Chicago Cubs and Los Angeles Dodgers, and in a nineteen-season career compiled a .264 batting average with 241 home runs and 775 RBI.

Tommy Lasorda, then a scout for the Los Angeles Dodgers, offered Monday twenty thousand dollars to sign with the Dodgers out of Santa Monica High School in 1963, but Arizona State University coach Bobby Winkles - who was also from Monday's native Arkansas - convinced Rick and his mother Nelda that he would look after Monday.

A star for the Sun Devils under head coach Winkles on a talented team that included freshman Reggie Jackson, Monday led the Sun Devils to the 1965 College World Series championship over Ohio State and earned All-America and College Player of the Year honors.

Monday was selected with the very first overall selection by Kansas City in the inaugural Major League First-Year Player Draft in 1965 and started his major league career with the Athletics. He then spent several productive years with the Chicago Cubs, and was traded to the Dodgers just in time to join a team that won the National League pennant in 1977 and 1978. Monday's finest season as a professional came in 1976 as a member of the Cubs. Batting in the leadoff position, he hit .272 and established career highs in home runs, runs, RBI, total bases, slugging percentage, and finished eighteenth in MVP voting.

The two most famous moments of Monday's career were associated with the Dodgers. The first occurred on April 25, 1976 during a game at Dodger Stadium. Two protesters, William Thomas and his eleven-year-old son, ran into the outfield and tried to set an American flag they had brought with them on fire. Monday, then playing with the Cubs, noticed they had placed the flag on the ground and were fumbling with matches and lighter fluid so he dashed over and grabbed the flag to thunderous cheers. He handed the flag to Los Angeles pitcher Doug Rau, and the ballpark police officers arrested the two intruders. When Rick came up to bat in the next half-inning he got a standing ovation from the crowd, and the big message board behind the left-field bleachers flashed the message, "RICK MONDAY... YOU MADE A GREAT PLAY..." He later said, "If you're going to burn the flag, don't do it around me. I've been to too many veterans' hospitals and seen too many broken bodies of guys who tried to protect it. I was angry when I saw them start to do something to the flag, and I'm glad I happened to be geographically close enough to do something about it. Whatever their protest was about, what they were attempting to do to the flag - which represents a lot of rights and freedoms that we all have - was wrong for a lot of reasons. Not only does it desecrate the flag, but it also desecrates the effort and the lives that have been laid down to protect those rights and freedoms for all of us. That feeling was very strongly reinforced by six years in the Marine Corps Reserves."

At the end of the season the Cubs traded Monday to the Dodgers in a five-player deal, and by 1981 he was mostly a utility player when the second moment occurred. In the deciding fifth game of the NLCS at Olympic Stadium in Montreal he smashed a ninth-inning home run off the Expos' Steve Rogers that proved to be the difference in a 2-1 Dodgers victory. Monday's home run dashed what turned out to be the Montreal's only chance at a pennant in their thirty-six-year history, and even today heartbroken Expos' fans refer to the fifth game of the NLCS as "Blue Monday." Los Angeles went on to win the 1981 World Series, defeating the New York Yankees four games to two.

Soon after his retirement as a player Monday became a broadcaster for the Dodgers, beginning in 1985 by hosting the pregame show and calling play-by-play on cable TV. Then from 1989 to 1992 he moved further south to call San Diego Padres games alongside fellow Marine Jerry Coleman.

Interestingly, Monday was born on the exact same day, month and year as Jay Johnstone, who was a fellow outfielder and teammate of his on the Dodgers' 1981 World Series champions. Both also served in the Marine Corps Reserve and played for the Cubs, Athletics and Dodgers.

On August 25, 2008 Monday was presented with an American flag flown over Valley Forge National Historical Park in honor of his 1976 rescue, and prior to the September 2, 2008 Los Angeles Dodgers game he was awarded a Peace One Earth medallion by Patricia Kennedy, founder of the non-profit organization 'Step Up 4 Vets,' in recognition of his actions in 1976 and his military service with the Marine Corps. The Baseball Hall of Fame also named Monday's quick-thinking act as one of the One Hundred Classic Moments in the history of the game.

Rick Monday is still in possession of the flag he saved from being burned - even though he was offered a million dollars for it. Along with the flag he has a copy of the 16-mm footage taken by a fan, Dodgers' broadcaster Vin Scully's play-by-play of the incident, and the now-famous photo by James Roarke of Monday just as he grabbed the flag (see Photo Gallery).

JIM MORA
"Playoffs?"

James Earnest Mora is a former head football who was born on May 24, 1935 in Glendale, California. He played football at Occidental College, and later coached the USFL's Philadelphia/Baltimore Stars and the NFL's New Orleans Saints and Indianapolis Colts.

As an NFL coach Mora is best known for three things: turning two of the NFL's worst franchises into perennial post-season contenders, his lack of success once he got his teams to the playoffs, and his often impassioned post-game tirades and press conferences, including his oft-quoted "Coulda, Woulda, Shoulda," "Diddly Poo," and "Playoffs?" tirades.

As a youth Mora became an Eagle Scout in 1950, and was presented the Distinguished Eagle Scout Award as an adult by the Boy Scouts of America. When Mora was a tight end at Occidental College his college roommate was Jack Kemp, who became an all star quarterback with the Buffalo Bills, a

U.S congressman for eighteen years, and a presidential candidate in 1988.

After playing three years of service football in the Marine Corps Mora became an assistant coach at his alma mater in 1960. He moved up to head coach of Occidental a few years later, and also coached for a number of years at Stanford, Colorado and Washington

In 1978 he moved to the professional ranks as the defensive line coach for the NFL's Seattle Seahawks under head coach Jack Patera. Then in 1983 the United States Football League came into existence, and Mora became head coach of the Philadelphia Stars. During his tenure with the team Mora compiled a 48-13-1 record, won two USFL championships, was named Coach of the Year in 1984, and was considered by many to be the best coach in the short history of the USFL. After the league folded in 1985 Mora took over as head coach of the NFL's New Orleans Saints.

After a 7-9 record his first season, Mora led the Saints to a 12-3 record in 1987. That was the Saints' first season over .500, and their first playoff appearance. In week six of that season the Saints lost a 24-22 game to the San Francisco 49ers, missing a last-second field goal. After the game Mora launched what became known as his "Coulda, Woulda, Shoulda" speech. In his postgame press conference Mora angrily said, "The Saints ain't good enough. We're close, and close don't mean shit. I'm tired of coming close. I'm pissed off right now. You bet your ass I am. I'm sick of coulda, woulda, shoulda, coming close, if only."

The Saints responded by winning their last nine games, and the team's 12-3 record was the second-best in the NFL that year. Unfortunately for them, the 49ers had the league's best record (13-2) and played in the same division. As a result the Saints were a wild-card team, and although they lost their first playoff game to the Minnesota Vikings Mora received the NFL Coach of the Year Award.

In 1991 Mora's Saints finished 11-5 and won the NFC West for the first time in team history, but the Saints lost to the Atlanta Falcons in the first round of the playoffs. Then in 1992 Mora led the Saints to their second twelve-win season in six years, finishing 12-4, but they were a wild card again because the 49ers finished 14-2 and for the second time in two years they were upset at home in the first round the playoffs by Philadelphia. This would be the Saints' last playoff game during Mora's tenure, leaving him with an 0–4 playoff record in New Orleans.

Mora served as a color analyst for NBC in 1997, and in 1998 replaced Lindy Infante as head coach of the Indianapolis Colts. The team struggled to a 3–13 mark in his first year with a rookie Peyton Manning learning the ropes at quarterback, but they had an amazing turnaround to 13-3 in 1999 thanks in large part to the addition of rookie running back Edgerrin James. After a first-round bye the Colts lost their first playoff game in the AFC Divisional Playoffs to the Tennessee Titans.

The Colts finished 10-6 in 2000 and made the playoffs once again, but the team lost a wild-card round playoff game to the Miami Dolphins by a score of 23-17 in overtime. This defeat dropped Mora's overall postseason record to 0-6. Coincidentally, just hours after Mora lost what would be the sixth and final playoff game of his career, his former team, the New Orleans Saints, won their first-ever playoff game.

On November 25, 2001, after a loss to the San Francisco 49ers which dropped the Colts to 4–6, Mora made his famous "Playoffs?" tirade. Of the Colts' performance Mora said, "Well, I'll start off by saying this… do not blame that game on the defense, okay? I don't care who you play -

whether it's a high school team, a junior college team, a college team - much less an NFL team. When you turn the ball over five times -four interceptions, one for a touchdown, three others in field position to set up touchdowns - you ain't going to beat anybody I just talked about. Anybody. All right? And that was a disgraceful performance in my opinion. We threw that game. We gave it away by doing that... pitiful!"

Later in the press conference, in response to a reporter's question about the Colts' chances for making the playoffs, Mora said in a high-pitched, incredulous sounding voice, "What's that? Ah... playoffs? Don't talk about... playoffs? You *kidding* me? Playoffs? I just hope we can win a game! Another *game*."

His "Playoffs" sound bite is regularly played back as a joke on a number of sports radio talk shows when discussing mediocre NFL teams or playoffs of other sports. The tirade would even go on to be featured in a Coors Light commercial in 2006 as part of an ad campaign that recreated NFL coaches' infamous press conferences with twenty-something male actors asking the coaches inane and unrelated non-football questions about the beer. In an appearance on *The Best Damn Sports Show* in 2003 Mora talked about going to autograph signings and having a kid ask him to say "playoffs" in lieu of an autograph.

Despite his solid regular season record (125–106), the biggest criticism of Jim Mora has been his NFL teams' inability to win any playoff games in six appearances. As of this writing he is a sports analyst on the NFL Network, primarily on the show *Point After,* and his son Jim L. Mora is the head coach of the Seattle Seahawks.

KEN NORTON
Going the Distance

Kenneth Howard Norton Sr. is a former multi-time world champion heavyweight boxer who was born August 9, 1943 in Jacksonville, Illinois. Norton was an outstanding athlete at Jacksonville High School, where his track coach once entered him in eight events and he placed first in them all. As a result the "Ken Norton Rule," which limits participation of an athlete to a maximum of three track and field events, was instituted in Illinois high school sports. After graduating from high school Norton went on to Northeast Missouri State University (now Truman State University) on a football scholarship.

Norton started boxing while he was serving in the Marine Corps from 1963-1967, compiling a 24-2 record en route to three All-Marine Heavyweight titles. He turned professional following the National AAU finals in 1967,

After suffering his first boxing defeat at the hands of Jose Luis Garcia in 1970 Norton was given the motivational book

Think and Grow Rich by Napoleon Hill, which he states in his autobiography, *Going the Distance*, changed his life. After reading it he went on a fourteen fight winning streak, including a shocking victory over Muhammad Ali in 1973 to win the North American Boxing Federation Heavyweight Champion title. In that bout Norton broke Ali's jaw, leading to only the second defeat for "The Greatest" in his career. Six months later Ali avenged the loss when he beat Norton in a split-decision.

In 1974 Norton fought George Foreman for the World Heavyweight Championship and was stopped in two rounds. In 1975 he regained the NABF Heavyweight Title when he defeated Jerry Quarry by TKO in the fifth round, and then avenged his above-mentioned 1970 loss to Jose Luis Garcia by knocking out Garcia in round five.

In 1976 Norton would again fight Ali, who was the World Heavyweight Champion since regaining the title with an eighth round knockout of George Foreman in 1974. Many observers have said this was the beginning of Ali's decline as a boxer. In one of the most disputed fights in history the bout was even on the judges' scorecards going into the final round - which Ali won on both the referee and judges' scorecards to retain the title. The judges scored the bout 8-7 for Ali, and the referee scored it 8-6 for Ali. The last time a heavyweight champion had lost the title by decision was Max Baer to Jim Braddock, and thirty-one years later Ali-Norton III was no exception. The January 1998 issue of *Boxing Monthly* listed Ali-Norton as the fifth most disputed title fight decision in boxing history.

Following the Leon Spinks upset of Muhammad Ali for the championship in February of 1978 Spinks elected to fight a return bout against Ali rather than face the still dangerous Norton, who was the top contender - and a fellow Marine.

The World Boxing Council, which had mandated a Spinks-Norton bout for their championship, then withdrew its recognition of Spinks as champion.

On March 18, 1978 the WBC made history by naming Norton its champion without benefit of a title match. The WBC stated that they considered Norton's victory over Jimmy Young, which was sanctioned by the WBC as a title eliminator, a retroactive championship match - but Norton wouldn't hold onto the title for very long.

In his first defense of the WBC title Norton and new number one contender Larry Holmes met in a classic fight, and after fifteen brutal rounds Holmes was awarded the title via a close split decision. The March 2001 edition of *Ring* magazine listed the final round of the Holmes-Norton bout as the seventh most exciting round in boxing history. Holmes-Norton is ranked as the tenth greatest heavyweight fight of all time by Monte D. Cox, a member of the International Boxing Research Organization (IBRO). Holmes went on to become the second longest reigning World Heavyweight Champion in the history of boxing behind Joe Louis.
He had.

Norton was past his prime when he retired in 1981 with a career professional record of 42-7-1, and is a 1989 inductee of the World Boxing Hall of Fame, a 1992 inductee of the International Boxing Hall Of Fame, a 2004 inductee into the United States Marine Corps Sports Hall of Fame, and a 2008 inductee into the WBC Hall of Fame. The 1998 holiday issue of *Ring* ranked Norton number twenty-two in "The Fifty Greatest Heavyweights of All Time." Norton received the Boxing Writers Association of America J. Niel Trophy for "Fighter of the Year" in 1977, and he received the "Napoleon Hill Award" for positive thinking in 1973.

Norton has appeared in approximately twenty motion pictures, and in fact the character of "Apollo Creed" in *Rocky* was initially going to be played by him prior to Carl Weathers being selected. He additionally worked as a television and radio sports commentator, and also appeared on some popular TV series such as *Knight Rider* and *The A-Team* opposite fellow Marine George Peppard.

Ken Norton was twice voted "Father of the Year" by the *Los Angeles Sentinel* and the *Los Angeles Times*. His son, Ken Norton Jr., played football for UCLA and in the NFL, and in tribute to his father's career would strike a boxing stance in the end zone and throw a punching combination at the goalpost pad each time he scored a defensive touchdown.

Norton also fathered a daughter, Kenisha, and two other sons, KeneJon and Keith, who became a sportscaster for Channel Two News in Houston, Texas after following his father's footsteps and serving in the Marine Corps.

Ken Norton now resides in San Clemente, California. His autobiography is titled *Going The Distance: The Ken Norton Story.*

BUM PHILLIPS
Houston Oilers

Oail Andrew "Bum" Phillips is a former NFL football coach who was born on September 29, 1923 in Nederland, Texas. Bum got his nickname from his older sister, who could not pronounce the word "brother."

Phillips played football at Lamar Junior College (now Lamar University) in Beaumont, Texas but enlisted in the Marine Corps shortly after the attack on Pearl Harbor plunged the United States into World War II. After he returned from the war he enrolled at Stephen F. Austin State University in Nacogdoches, Texas where he lettered in football and graduated with a degree in Education in 1949.

His college coaching stints included serving as an assistant coach at Texas A&M University, the University of Houston, Southern Methodist University (under fellow Marine Hayden Fry), the University of Texas at El Paso, and Oklahoma State University.

In the early 1970s Phillips went to the NFL when he was hired by Sid Gillman to serve as a defensive assistant coach for the San Diego Chargers. A few years later Gillman became head coach of the Houston Oilers and brought Phillips with him as his defensive coordinator.

In 1975 Phillips was named head coach and general manager of the Oilers and served in that capacity through 1980. He became the winningest coach in franchise history (59-38 record), and was known for his folksy mannerisms, and for wearing his trademark cowboy hat on the sidelines - except when the Oilers played in the Astrodome or other domed stadiums. (He stated that his mother taught him not to wear a hat indoors.) Under Phillips the Oilers reached the AFC Championship Game in two consecutive seasons, losing to the eventual Super Bowl-champion Steelers 34-5 in 1978 and 27-13 in 1979. Both teams were members of the competitive AFC Central Division and thus played three times in both 1978 and 1979, fueling a monumental rivalry. During this period of league-wide AFC dominance some commentators actually considered Houston and Pittsburgh to be the two best teams in the entire NFL. As Phillips remarked at the time, "The road to the Super Bowl goes through Pittsburgh."

Phillips later worked as a football color analyst for television and radio, and has since retired to his horse ranch in Goliad, Texas. As of this writing his son Wade is head coach of the Dallas Cowboys.

BARNEY ROSS
Celebrity Hero

Barney Ross, (December 23, 1909 - January 17, 1967) born Dov-Ber Rasofsky, was a professional boxer. After his beloved father, who was a rabbi, died in his arms after being shot during a robbery Ross, a rabbinical student, lost his faith in God and abandoned his studies. In an amazing life he became a street brawler alongside his buddy Jack Ruby, went to work for Al Capone, transformed himself into the first three-weight-class champion in boxing history, became a leader for the Jewish people and all Americans in the battle against Hitler and Nazism, and then gave up boxing and insisted on fighting as a Marine in World War II where at the age of thirty-three he single-handedly killed twenty-two Japanese soldiers during a battle on Guadalcanal.

Ross was perhaps the first Jewish-American person ever to have a ticker-tape parade in his honor, and was likely the first to be celebrated by a president when Franklin D. Roosevelt honored him in a Rose Garden ceremony after his WW II heroics. As a professional athlete he made more money than Babe Ruth, and had his own namesake candy bar too. The legend of Barney Ross is pure America from his early struggles to legendary heroism to heartbreak… and his friend Jack Ruby would of course make history in his own way by killing Lee Harvey Oswald.

Young Dov-Ber grew up on Chicago's mean streets, ultimately ignoring his father's admonition that Jews do not fight back. "Let the atheists be the fighters," Ross later recalled being told by his father. "The 'trumbeniks,' the murderers - we are the scholars." Ross's ambition in life was to become a Jewish teacher and a Talmudic scholar, but his life was changed forever when his father was shot dead resisting a robbery at his small grocery store. Prostrate from grief, his mother Sarah suffered a nervous breakdown and his younger siblings - Ida, Sam and George - were placed in an orphanage or farmed out to other members of the extended family. Dov was left to his own devices.

In the wake of the tragedy Dov became vindictive towards everything and turned his back on the orthodox religion of his father. He began running around with local toughs and developed into a street brawler, thief and money runner employed by Al Capone. Dov's goal was to earn enough money to buy a home so that he could reunite his family, and when he decided boxing was the vehicle to get it he began training with his friend Jack Ruby.

After winning amateur bouts Dov would pawn the awards - things such as watches - and set the money aside for his family. There is speculation that Al Capone bought up tickets to his early fights, knowing some of that money would be funneled to Dov. Plagued by his father's death and feeling an obligation not to sully his name, Dov Rasofsky took the new name "Barney Ross." The name change was also part of a larger trend by Jews to assimilate themselves in the U.S. by taking American-sounding names. Strong, fast and possessed of a powerful will, Ross was soon a Golden Gloves champion and went on to dominate the lighter divisions as a pro.

During that period rising Nazi leader Adolf Hitler was using propaganda to spread his virulently anti-Jewish philosophy, and Ross was seen by American Jews as one of their greatest advocates. He represented the concept of Jews finally fighting back. Idolized and respected by all Americans, Ross showed that Jews could thrive in their new country and he made his stand against Hitler and Nazi Germany a public one. He knew that by winning boxing matches he was displaying a new kind of strength for Jews. He also understood that Americans loved their sports heroes, and if Jews wanted to be embraced in the U.S. they would have to assume such places in society. Even though Ross had lost faith in religion, he openly embraced his role as a leader of his oppressed people.

Ross occupies a rarified place as one of boxing's few triple division champions - lightweight, junior welterweight and welterweight. He was never knocked out in eighty-one fights, and held his title against some of the best competition in the history of the divisions, defeating great Hall of Fame champions like Jimmy McLarnin and Tony Canzoneri in epic battles that drew crowds of more than fifty thousand.

His first paid fight was on September 1, 1929 when he beat Ramon Lugo by a decision in six rounds. After ten wins in a row he lost for the first time to Carlos Garcia on a decision in ten. Then over the next thirty-five bouts his record was 32-1-2, including a win over former world champion Bat Battalino. Another bout included former world champion Cameron Welter. Then, on March 26, 1933 Ross was given his first shot at a world title when he faced world Lightweight and Jr. Welterweight champion and fellow three division world champion Tony Canzoneri in Chicago. In one night Ross became a two division world champion when he beat Canzoneri by a decision in ten rounds.

In his last fight Ross defended his title on May 31, 1938 against fellow three division world champion Henry Armstrong, who beat him by a decision in fifteen. Although Armstrong pounded him inexorably and his trainers begged Ross to let them stop the fight he absorbed the abuse and refused to stop or go down. Barney Ross had never been knocked out in his career, and he was determined to leave the ring on his feet. Some boxing experts view Ross's performance against Armstrong as one of the most courageous in history, and believe his will to survive every tough fight on his feet had to do with his understanding of his symbolic stature as a Jew - that is, Jews would not only fight back, but they wouldn't go down.

Ross retired with a record of seventy-two wins, four losses, three draws and two no-contests, with twenty-two wins by way of knockout, and was ranked number twenty-one on *Ring Magazine's* list of the Eighty Best Fighters of the Last Eighty Years.

While in retirement and in his early thirties Ross decided to fight in World War II and joined the Marine Corps. The Marines wanted to keep him stateside and use his celebrity status to boost morale, like many athletes of the era such as heavyweight champion Jack Dempsey, but Ross insisted on fighting for his country. He was sent to Guadalcanal in the South Pacific, where one night he and three comrades were trapped under enemy fire. All three of his fellow Marines were wounded, as was Ross, and he was the only one able to fight. And fight he did. Ross gathered his comrades' rifles and grenades and single-handedly fought nearly two dozen Japanese soldiers over an entire night, killing them all by morning. Two of the Marines with him had died in the battle, but he carried the remaining man (who weighed 230 pounds compared to Ross' 140) to safety on his shoulders. Because of his heroism Barney Ross was awarded America's third highest military honor, the Silver Star, as well as a Presidential Citation. As Americas greatest "celebrity war hero" he was honored by President Roosevelt in a Rose Garden ceremony.

During his time on Guadalcanal Ross began a lifelong friendship with the famous Father Frederic Gehring, a chaplain who wrote regular correspondences for Reader's Digest magazine and was known as the 'Padre of Guadalcanal.' Gehring considered Ross a national treasure who defied logic when it came to bravery and the defense of principle, and one Christmas Eve before Barney and his Marines were to go to battle Gehring asked Ross to take part in what would become one of the most poignant such events of the war. Barney was the only one capable of playing a temperamental organ on the tropical island, so Gehring asked him to learn *Silent Night* and other Christmas songs for the troops. Ross played those songs and sang with the homesick young men, after which Gehring implored him to play a Jewish song. He chose a melancholy song called "My Yiddishe Momma" about a child's love for his self-sacrificing mother. Many of the Marines knew the melody of the song because Ross always had it played when he entered the ring, but when the Marines heard the heart-rending lyrics newspaper reports said they were all in tears.

After Barney Ross' single-handed victory in the battle at Guadalcanal he was viewed as almost superhuman, particularly after all he'd had to overcome during his troubled life, but during his recovery from the wounds suffered in that battle he developed an addiction to the morphine administered for pain. Back in the states the morphine became heroin, and the habit became so bad he would sometimes spend five hundred dollars a day on the drug. Ross went to a recovery center and beat his addiction, and later gave lectures to high school students about the dangers of drugs.

Ross spent his last days using his celebrity status in promotional work for casinos and other businesses and remained with second wife Cathy Howlett, although they never had children. Happy to have reached the goals he had set to reunite his family and become a world champion in boxing, he wrote an autobiography titled *No Man Stands Alone*. He also remained loyal to his friend Jack Ruby, and testified as a character witness on Ruby's behalf at his trial for killing Lee Harvey Oswald, the alleged assassin of President John F. Kennedy.

Barney Ross died in his hometown of Chicago when he was fifty-seven years old.

TOM SEAVER
Tom Terrific

George Thomas "Tom" Seaver is a former Major League Baseball pitcher who was born on November 17, 1944 in Fresno, California to Betty Lee Cline and Charles Henry Seaver. He broke into the major leagues in 1967, retired in 1987, and played for four different teams during his career - but he is primarily associated with the New York Mets. As the Mets' all-time leader in wins, Seaver is considered the greatest player in that team's history, as well as one of the best starting pitchers in the history of baseball.

Nicknamed "Tom Terrific" and "The Franchise," he had 311 wins, 3,640 strikeouts, and a 2.86 ERA during a twenty-year career, won the National League Rookie of the Year Award in 1967, three NL Cy Young Awards as the league's best pitcher, and in 1992 he was inducted into the Baseball Hall of Fame.

Seaver joined the Marine Corps Reserve on June 28, 1962, served with AIRFMFPAC at 29 Palms, California through July 1963, and performed six months of active duty while enrolled at Fresno City College. By then he was much stronger and threw with greater velocity than he had in high school but still had the same fine control of his pitches. In anticipation of the following season he was being recruited to pitch for the University of Southern California and was sent to pitch for the Alaska Goldpanners to see if he was worthy of a scholarship. After a stellar season in which he pitched and won a game in the national tournament with a grand slam he was accepted for a USC scholarship, and as a sophomore there Seaver posted a 10-2 record. In 1966 he signed a contract with the Atlanta Braves, who had drafted him number one, but the contract was voided by Baseball Commissioner William Eckert because his college team had played two exhibition games. Seaver decided to finish the college season but because he had signed a pro-contract the NCAA ruled him ineligible, and after his father complained to Eckert about the unfairness of the situation and threatened a lawsuit Eckert ruled that other teams could match the Braves' offer. The Mets were subsequently awarded his signing rights in a lottery drawing among the three teams willing to match the Brave" terms.

Seaver spent one season with the Jacksonville Suns of the International League and then joined New York in 1967. He won sixteen games for the last-place Mets, with eighteen complete games, 170 strikeouts and a 2.76 ERA - all Mets' records to that point - and was named the National League Rookie of the Year. He also played in the 1967 All-Star Game and got the save by pitching a scoreless fifteenth inning. In 1968 he won sixteen games again and recorded over two hundred strikeouts for the first of nine consecutive seasons, but the Mets moved up only one spot in the standings to ninth.

Then in 1969 Seaver and the "Miracle Mets" completed a remarkable season, coming from the depths of the National League to win their first World Series championship. Seaver won a league-high twenty-five games and his first Cy Young Award, and at year's end was presented with both the Hickok Belt as the top professional athlete of the year and *Sports Illustrated Magazine's* "Sportsman of the Year" award.

Seaver had four more twenty-win seasons and won two more Cy Young Awards in1973 and 1975, although 1971 was arguably his finest season when he led the league in ERA (1.76) and strikeouts (289 in 286 innings) while going 20-10. Between 1970 and 1976 he led the National League in strikeouts in five of the seven seasons, and finished second in 1972 and third in 1974. He also won three ERA titles as a Met. A famous quote about Seaver is attributed to Reggie Jackson was, "Blind men come to the park just to hear him pitch."

In what New York sports reporters dubbed the "Midnight Massacre," the Mets traded Seaver to the Cincinnati Reds on June 15, 1977 for Pat Zachary, Steve Henderson, Doug Flynn, and Dan Norman. He finished the 1977 season with twenty-one wins by going 14-3 with Cincinnati, including an emotional win over the Mets in his return to Shea Stadium. His departure from New York sparked sustained negative fan reaction as the Mets once again became the National League's worst team.

There was one milestone that had eluded him up to that point, and after having thrown five one-hitters for New York, including three no-hitters that were broken up in the ninth inning, Seaver finally recorded a 4-0 no-hitter against the St.

Louis Cardinals on June 16, 1978 at Riverfront Stadium. It was the only no-hitter of his professional career.

After the 1982 season Seaver was traded back to the Mets and had high expectations going into 1984 and the intention of finishing his career where he started it, so he and the Mets were stunned when he was claimed in a free-agent compensation draft by the Chicago White Sox. The Mets had incorrectly assumed that no one would pursue a high-salaried, thirty-nine-year-old starting pitcher and left him off the protected list. Faced with either reporting to the White Sox or retiring, Seaver chose the former.

Seaver pitched two and a half seasons in Chicago, and after his 298th win a reporter pointed out to White Sox catcher Carlton Fisk that following his upcoming start in Boston Seaver's next scheduled start would be in New York, and that the possibility existed that he might achieve a milestone there. Fisk emphatically stated that Seaver *would* win in Boston, and then would win his three hundredth.

On August 4, 1985 Seaver did just that in New York when he pitched a complete game against the Yankees on "Phil Rizzuto Day" which is significant because Seaver would later become Rizzuto's broadcast partner for Yankee games. It was also the same day fellow Marine Rod Carew, Seaver's 1967 American League Rookie of the Year counterpart, collected his three thousandth hit. Lindsey Nelson, a Mets radio and TV announcer during Seaver's salad days, called the final out for Yankees' TV flagship WPIX.

Seaver last season was in 1986 with the Boston Red Sox. At the time of his retirement he was third on the all-time strikeout list with 3,640, trailing only Nolan Ryan and Steve Carlton, and his lifetime ERA of 2.86 was third among starting pitchers in the "live-ball" era behind only Whitey Ford and Sandy Koufax.

A knee injury prevented him from appearing against the Mets in the 1986 World Series, but Seaver received the loudest ovations during player introductions prior to Game One. The Mets retired Seaver's uniform number forty-one in 1988, and as of 2007 he remains the only Met player to have his uniform number retired. Casey Stengel and fellow Marine Gil Hodges had their numbers retired as Met managers, and Jackie Robinson, who was never affiliated with the Mets, had his number retired by all teams.

LEON SPINKS
Montreal Mauler

Leon Spinks is a former boxer who was born on July 11, 1953 in St. Louis, Missouri. He had an overall record of twenty-six wins, seventeen losses and three draws as a professional, with fourteen knockout wins. While still an amateur he became a member of the United States Marine Corps, and later went from being heavyweight champion of the world to being homeless in little more than a decade.

Spinks won the gold medal in the light heavyweight division during the 1976 Summer Olympics in Montreal alongside brother Michael Spinks, who also won a gold medal in those games. Two years earlier, at the inaugural 1974 World Amateur Boxing Championships in Havana, Cuba he captured the bronze medal. His Olympic teammates included Sugar Ray Leonard, Leo Randolph and Howard Davis Jr.

Spinks debuted professionally on January 15, 1977 in Las Vegas and beat Bob Smith by a knockout in five rounds. His next fight, which was his debut abroad, was in Liverpool, England where he beat Scotty Child by a knockout in the first round. A couple of fights later he saw a slight improvement in opposition quality when he fought Pedro Agosto of Puerto Rico and knocked him out in the first, and he then drew with Scott LeDoux and beat Italian champion Alfio Riguetti by decision.

After that Spinks was ranked number one among the world's heavyweight challengers and made history in only his eighth fight when he won the heavyweight title from an aging and out-of-shape Muhammad Ali in a fifteen-round decision in Las Vegas on February 15, 1978 in what was the fastest ascent in history. Ali, who had not been the same since his last fight with Joe Frazier, expected an easy fight but was out-hustled by Spinks, who never seemed to tire. The victory over Ali was the peak of Spinks' career, and his iconic gap-toothed grin was featured on the cover of the February 19, 1978 issue of *Sports Illustrated* - but he would never again fight as efficiently because he was a party animal with a large entourage during his reign as heavyweight champion. As an interesting side note, a young Mr. T served as one of his bodyguards during that time.

Spinks was stripped of his world title by the WBC for refusing to defend it against their top-ranked contender, fellow Marine Ken Norton. Spinks instead agreed to fight a return bout against Ali for the WBA crown, and the WBC subsequently named Norton its champion - marking the first time a boxer had been awarded the heavyweight title without winning it in the ring. Because of the WBC's action Spinks was the last undisputed heavyweight champion until the emergence of Mike Tyson.

The second Ali fight was different from the first. Spinks lost the title at the Louisiana Superdome in New Orleans on September 15, 1978 in a unanimous fifteen-round decision. This time it was he who did not train hard enough and Ali, despite his declining skills, who gave every ounce of himself. By regaining the title, Ali became the first three-time heavyweight champion.

In his next fight Spinks went to Monte Carlo where he was knocked out in the first round by future WBA world heavyweight champion Gerrie Coetzee, and in 1980 he beat former world title challenger Alfredo Evangelista by a knockout in five, boxed to a draw in ten with Eddie "The Animal" Lopez, and beat the WBC's number one ranked challenger Bernardo Mercado by a knockout in nine.

After the win over Mercado Spinks challenged for the WBC world heavyweight championship once again. In his only fight of 1981 he faced Larry Holmes in Detroit and was knocked out in the third round. Then in 1982 Spinks decided to go down in weight and compete in the Cruiserweight division where he beat fringe contender Ivy Brown by a decision in ten rounds, and former and future title challenger Jesse Burnett by decision in twelve.

Spinks boxed on and off from 1981 to 1985 in both the heavyweight and cruiserweight divisions. When his brother Michael beat Holmes on September 21, 1985 the Spinks' became the first pair of brothers to have been world Heavyweight champions.

Spinks also performed in several boxer vs. wrestler matches in New Japan Pro Wrestling in the 1980s, including losing by submission to Antonio Inoki. In 1986 he lost to the WBA's world cruiserweight champion Dwight Muhammad Qawi by a knockout in six rounds, his second attempt at being a two-time world champion.

Leon Spinks boxed for eight more years with mixed results. In 1994 he lost a bout by knockout to John Carlo, and it was noteworthy for being the first time a former heavyweight champion had ever lost to a boxer making his pro debut. This humiliation was amplified by the fact Spinks lost by KO and was unable to land even a single punch. Leon Spinks finally retired at the age of forty-two after losing by decision in eight to Fred Houpe in 1995.

RICHARD STEELE
Third Man in the Ring

Richard Steele is a former boxing referee who was born in Los Angeles, California in 1944. He began his career as an amateur boxer while serving in the Marine Corps, compiling a record of twelve wins and three losses, and was the USMC Middleweight Boxing Champion from 1962 to 1963 while a teammate of future world Heavyweight champion Ken Norton. After his discharge he launched a professional career and compiled a record of sixteen wins and four defeats.

Steele began referring fights in the 1970s and he went on to referee 167 world title fights around the world. In 1983 he handled his first major fight, in which Aaron Pryor knocked out Alexis Arguello in ten rounds in their rematch. Other fights Steele refereed included Marvin Hagler vs. Thomas Hearns, Julio Cesar Chavez vs. Meldrick Taylor I, and the fight where Sugar Ray Leonard made his comeback after a three year lay-off in 1987 and beat Hagler.

Steele was sometimes involved in controversy, but none bigger than the one which occurred after Chavez-Taylor I. With Taylor ahead on the scorecards and seconds away from handing Chavez his first defeat he was dropped by a punch to the chin. He got up on the five count, but Steele decided to stop the fight with two seconds left in the last round. Taylor was clearly shaken and did not respond when Steele asked him, "Are you alive, or just breathing?" This defeat affected Taylor greatly and proved to be the beginning of the end of his career as a professional boxer. Many fans who saw the fight still argue as to whether or not the fight should have been stopped, considering the very short time left in the bout. Steele defended himself by declaring he was just trying to protect Taylor from more punishment and did not know how much time was left in the fight, saying, "No fight is worth a man's life." What many forget is Taylor had taken a severe battering. Doctor Flip Hopansky of the Nevada State Athletic Commission examined the young fighter and said he had a facial fracture, was urinating pure blood, and his face was grotesque. It was later estimated that Taylor had swallowed approximately two pints of blood during the course of the bout.

Outside the boxing ring Richard Steele has made a name for himself as a community conscious person, opening a gym - the Richard Steele Boxing Center - in Las Vegas, and helping out with Salvation Army charities. In 1999 he was given an award by South African President Nelson Mandela for refusing to referee fights in South Africa while the apartheid laws were still in use there.

LEE TREVINO
Super Mex

Lee Buck Trevino is a professional golfer who was born December 1, 1939 in Dallas, Texas into a family of Mexican ancestry. He is an icon for Mexican Americans, and is often referred to as "The Merry Mex" and "Super Mex."

He was raised by his mother Juanita and his grandfather Joe, who was a gravedigger. Trevino never knew his father Joseph, who left home when his son was small. Lee's childhood consisted of occasionally attending school and working to earn money for the family, and at age five he started working in the cotton fields.

Trevino was introduced to golf when his uncle gave him a few golf balls and an old club. He spent his free time sneaking into nearby country clubs to practice, and began as a caddy at the Dallas Athletic Club. Lee left school at fourteen to begin caddying full time, and earned thirty dollars a week as a caddy and shoe shiner. He was also able to practice golf since the caddies had three short holes behind their shack, and every day after work he would hit at least three hundred balls.

When Trevino turned seventeen he enlisted in the Marine Corps and served four years. Part of his time was spent playing golf with Marine Corps officers, and he claims being a golf partner helped earn him promotion to lance corporal.

After his discharge Trevino became a club professional in El Paso, Texas and made extra money gambling for stakes in head-to-head matches. He began play on the PGA Tour in 1967, and in his second U.S. Open golf championship shot 283 to come in eight shots behind champion Jack Nicklaus and earned $6,000 for finishing fifth. He won $26,472 as a rookie, forty-fifth on the PGA Tour money list, and was named Rookie of the Year by *Golf Digest*.

In 1968, his second year on the circuit, Trevino won the U.S. Open at the Oak Hill Country Club in Rochester, New York. During his career he won a total of twenty-nine times on the PGA Tour, including six majors. He was at his best in the 1970s, when he was Jack Nicklaus' chief rival. He won the money list title in 1970, and had ten wins in 1971 and 1972. These included the 1971 U.S. Open, which he took in an eighteen-hole playoff over Nicklaus. Two weeks later he won the Canadian Open, and the following week The Open Championship, becoming the first player to win those national titles in the same year. Trevino was awarded the Hickok Belt as the top professional athlete of 1971, *Sports Illustrated* magazine's "Sportsman of the Year," and was named the ABC Wide World of Sports Athlete of the Year.

Trevino was struck by lightning at the 1975 Western Open, suffered injuries to his spine, and underwent surgery to remove a damaged spinal disk - but back problems continued to hamper his play. Even so he was ranked second in McCormack's World Golf Rankings in 1980 behind Tom Watson and won his sixth major, the PGA Championship, at the age of forty-four.

Lee Trevino has also won more than twenty international and unofficial professional tournaments, and was one of the charismatic stars who were instrumental in making the Senior PGA Tour (now the Champions Tour) an early success. He claimed twenty-nine wins on that tour, including four senior majors, and topped the seniors' money list in 1990 and 1992.

GENE TUNNEY
A Man Must Fight

James Joseph "Gene" Tunney (May 25, 1897 - November 7, 1978) was the world heavyweight boxing champion from 1926-1928 who defeated Jack Dempsey twice, first in 1926 and then in 1927. Tunney's successful title defense against Dempsey is one of the most famous bouts in boxing history, and is known as "The Long Count Fight." Tunney later retired as an undefeated heavyweight after his victory over Tom Heeney in 1928.

Tunney was regarded as an extremely skillful boxer who excelled in defense. In addition to beating Dempsey, the most famous fighter of his era, he defeated Tommy Gibbons, Georges Carpentier, and many other fine boxers. When World War I broke out Tunney enlisted in the Marines and continued to box, eventually earning the Light Heavyweight Championship of the American Expeditionary Forces. Upon returning from France he boxed as a Light Heavyweight, taking on Soldier Jones, Battling Levinsky, and Jack Burke. Tunney also had a brief acting career, starring in the movie *The Fighting Marine* in 1926. Unfortunately, no prints of this film are known to exist.

He was named *Ring Magazine's* first-ever Fighter of the Year in 1928, and was later elected to the World Boxing Hall of Fame in 1980, the International Boxing Hall of Fame in 1990, and the United States Marine Corps Sports Hall of Fame in 2001.

In 1928 Tunney married a wealthy socialite, the former Mary "Polly" Lauder, and fathered four children, among them John V. Tunney (born 1934), who was a U.S. Representative and Senator from California from 1965 until 1977.

Mrs. Tunney's grandfather was George Lauder, a first cousin and business partner of industrialist and philanthropist Andrew Carnegie, founder and head of the Carnegie Steel Company of Pittsburgh, Pennsylvania. Her father, George Lauder, Jr., was a philanthropist and yachtsman whose 136-foot schooner once held the record for the fastest trans-Atlantic yacht passage ever made. According to a 2007 biography Gene promised Polly that he would quit boxing, and defended his title only one more time against Tom Heeney of New Zealand.

Tunney was a thinking fighter who preferred to make a boxing match into a game of chess, which was not popular during times when such sluggers as Jack Dempsey, Harry Greb and Mickey Walker were commanding center stage. Tunney's style was influenced by other noted boxing thinkers such as James J. Corbett and Benny Leonard, but it is incorrect to think of him as only a fancy dan, stick and move fighter in the Ali style. While Tunney's heavyweight fights against Gibbons, Carpentier, and Dempsey were noted for his fleet-footed movement and rapid fire jabbing his earlier fights, especially the five fights against Harry Greb, were noted for Tunney's vicious body punching and willingness to slug it out toe-to-toe. It was Benny Leonard who advised Tunney the only way to beat Harry "The Human Windmill" Greb was to aim punches at Greb's body rather than his elusive and often butting head.

Always moving and boxing behind a good solid left jab, Tunney would start reading and dissecting his opponent from the first bell, preferring to stay on the outside to nullify any attack and using quick counters to keep them off balance. Not known as a big puncher, Tunney could hit with venom if need be, especially once he had figured his opponents out or they were exhausted and hurt.

In his fights against Jack Dempsey today's viewer can see Tunney's style. Hands held low for greater power, fast footwork that adjusted to every move his opponent made, and quick, accurate one-two style counter-punches.

Tunney, while not known for having one of the truly great chins in the history of boxing, did own a solid one. He was never knocked out, and the only time he was ever knocked down was in the second fight with Dempsey during the infamous Long Count.

In 1932 Tunney published a book called *A Man Must Fight*, in which he gave comments on his career and boxing techniques. Gene Tunney died in 1978 at the age of eighty-one and was interred at Long Ridge Union Cemetery in Stamford, Connecticut.

BILL VEECK
Veeck, As In "Wreck"

William Louis Veeck, Jr. (February 9, 1914 - January 2, 1986), also known as "Sport Shirt Bill," was a native of Chicago, Illinois and a pioneering Major League Baseball franchise owner and promoter. He was best known for his flamboyant publicity stunts and the innovations he brought to the league during his ownership of the Cleveland Indians, St. Louis Browns and Chicago White Sox. Veeck was the last owner to purchase a baseball franchise without an independent fortune, and is responsible for many significant and contributions to baseball.

In response to his critics Veeck once said, "All I ever said is that you can draw more people with a losing team, plus bread and circuses, than with a losing team and a long, still silence."

While Veeck was growing up in Hinsdale, Illinois his father, William Veeck, Sr., became president of the Chicago

Cubs. Veeck Sr. was a local sportswriter who had written several columns about what he'd do differently if he ran the Cubs - and the team's owner, William Wrigley Jr., took him up on it. Growing up in the business, young Bill Veeck worked as a vendor, ticket seller and junior groundskeeper, and when his father died in 1933 he left Kenyon College and eventually became club treasurer for the Cubs. In 1937 Veeck planted the famous ivy on the outfield wall at Wrigley Field and is responsible for the construction of the hand-operated centerfield scoreboard still used there.

In 1941 Bill left Chicago and purchased the American Association Milwaukee Brewers in a partnership with former Cubs star and manager Charlie Grimm, and after winning three pennants in five years he sold the franchise in 1945 for a $275,000 profit.

While a half-owner of the Brewers Veeck served for nearly three years in the Marines during World War II in an artillery unit. During this time a recoiling artillery piece crushed his leg, requiring amputation first of the foot, and shortly thereafter of the leg above the knee. Over the course of his life he had thirty-six operations on the leg, a series of wooden legs, and as an inveterate smoker cut holes in them to use as an ashtray.

According to Veeck's memoirs in 1942, before entering the military he acquired backing to purchase the financially strapped Philadelphia Phillies and planned to stock the club with stars from the Negro Leagues. He then claimed that Commissioner Kenesaw Mountain Landis, a virulent racist, vetoed the sale and arranged for the National League to take over the team. Although this story has long been part of accepted baseball lore, in recent years its accuracy has been challenged by some researchers.

329

In 1946 Veeck finally became the owner of a major league team, the Cleveland Indians, by using a debenture-common stock group making remuneration to his partner's non-taxable loan payments instead of taxable income. He immediately put the team's games on radio and set about putting his own indelible stamp on the franchise.

In 1947 he signed Larry Doby, who became the first African-American player in the American League, and followed that one year later by inking Satchel Paige to a contract which made the hurler the oldest rookie in major league history. There was much speculation at the time about Paige's true age, with most sources stating that he was forty-two when he joined the Indians. Many sports writers mocked Veeck's decision. One wrote that if Paige had been white no one would have thought to sign him, and Veeck countered, "If Satchel had been white, he would have been in the majors twenty years ago."

As in Milwaukee, Veeck took a whimsical approach to promotions, hiring rubber-faced Max Patkin, the "Clown Prince of Baseball'" as a coach. Patkin's appearance in the coaching box delighted fans and infuriated the front office of the American League.

Although Veeck had become extremely well-liked, an attempt in 1947 to trade popular player-manager Lou Boudreau to the St. Louis Browns led to mass protests and petitions. In response Veeck visited every bar in Cleveland to apologize for his mistake and reassured fans that the trade would not occur. The following year, led by Boudreau's .355 batting average, Cleveland won its first pennant and World Series since 1920 - but in 1949 Veeck's first wife divorced him and since most of his money was tied up in the Indians he was forced to sell the team to fund the divorce settlement.

After marrying Mary Frances Ackerman, Veeck bought an eighty percent stake in the St. Louis Browns in 1951. Hoping to force the Cardinals out of town, Veeck spited Cardinals owner Fred Saigh by hiring Cardinal greats Rogers Hornsby and Marty Marion as managers and Dizzy Dean as an announcer. It didn't work.

Some of Veeck's most memorable publicity stunts occurred during his tenure with the Browns, including the famous appearance on August 19, 1951 by midget Eddie Gaedel for which Veeck predicted he would be most remembered.

Veeck considered moving the Browns back to Milwaukee (where they had played their inaugural season in 1901), but was denied permission by the other American League owners. He also wanted to move his club to the lucrative Los Angeles market, but was denied again. He then got in touch with a group that was looking to bring big-league ball to Baltimore, but the owners voted that move down as well. Realizing the other owners simply wanted him out of the picture, Veeck then agreed to sell his entire stake to a group which then moved the team to Baltimore as the Orioles.

In 1959 Veeck became head of a group that purchased a controlling interest in the Chicago White Sox, who went on to win their first pennant in forty years and broke a team attendance record for home games with 1.4 million. The next year the team broke the same record with 1.6 million visitors to Comiskey Park with the addition of the first "exploding scoreboard" in the major leagues which produced sound effects and shot fireworks whenever the White Sox hit a home run. According to Lee Allen in *The American League Story*, after the Yankees watched the exploding scoreboard a few times Clete Boyer, their weak-hitting third baseman, hit the ball over the outfield fence and Mickey Mantle and several other Yankee players came out of the dugout waving sparklers. The point was not lost on Veeck.

Veeck wasn't heard from again in baseball circles until 1975 when he returned as owner of the White Sox. His return rankled baseball's owner establishment, with most of the old guard viewing him as a pariah for exposing most of his peers in his 1961 book *Veeck As In Wreck*, and for testifying against the reserve clause in the Curt Flood case.

Veeck presented a Bicentennial-themed "Spirit of '76" parade on opening day in 1976, casting himself as the peg-legged fifer bringing up the rear. In the same year he reactivated long-retired Minnie Miñoso for eight at-bats in order to give Miñoso a claim towards playing in four decades, and did so again in 1980 to expand the claim to five.

The 1979 season was arguably Veeck's most colorful and controversial. On April 10th he offered fans free admission the day after a 10-2 Opening Day shellacking by the Toronto Blue Jays, and then on July 12th, with an assist from son Mike and radio host Steve Dahl, held one of his most infamous promotion nights, Disco Demolition Night, which resulted in a riot at Comiskey Park and a forfeit to the visiting Tigers.

Finding himself no longer able to financially compete in the free agent era, Veeck sold the White Sox in January of 1981 and retired to his home in St. Michaels, Maryland near where he had earlier discovered White Sox star Harold Baines while Baines was in high school there.

Bill Veeck, weak from emphysema and having had a cancerous lung removed in 1984, died in 1996 of a pulmonary embolism at age seventy-one. He was elected five years later to the Baseball Hall of Fame.

MIKE WEAVER
Hercules

Michael Dwayne Weaver is a former boxer who was born on July 7, 1952 in Gatesville, Texas. Weaver was a member of the United States Marine Corps from 1968 to 1971, and while in Vietnam during that time got into amateur boxing and training. He notably fought Duane Bobick, a future amateur star out of the Navy, and in a fight where both men were down Weaver was outpointed.

By 1972 Weaver was living and training in California and had taken up professional boxing. In his early career he was considered a journeyman opponent and was frequently brought in on short notice and overmatched against more experienced and developed contenders. Weaver was also used as a sparring partner for Muhammad Ali and fellow Marine Ken Norton, who famously nicknamed him "Hercules."

After a few losses early on to tough fringe contenders like Howard Smith and Larry Frazier, Weaver showed signs of improvement. He fought both Bobick brothers, losing a debatable ten round decision to Rodney, and was stopped on a cut in the seventh against old amateur rival Duane.

In 1976 Weaver beat well regarded veteran Jody Ballard, and in 1978 lost two close decisions to Stan Ward for the California State Heavyweight title and to Leroy Jones for the NABF heavyweight title.

Then in late 1978 Weaver got a new team and manager and reeled off five straight knockouts, two of which came over top ranked opponents. In October of 1978 he came off the floor to knock out hard hitting Colombian Bernardo Mercado in five, and in January of 1979 knocked out old foe Stan Ward in nine to win the USBA heavyweight title.

These wins got him a high profile title fight with reigning and undefeated WBC champion Larry Holmes in New York's Madison Square Garden in June of 1979. Then-new cable channel HBO bought the rights to the fight, since Weaver was so lowly-regarded that the fight was seen as a mismatch and the networks didn't want anything to do with it (Weaver was 20-8 to Holmes' 30-0).

Weaver proved to be better than expected however, and gave Holmes a tough fight. In the end Holmes would rally, knocking Weaver down with an uppercut in the eleventh, and stopping him on his feet in the twelfth. Although Weaver lost his surprise showing made him a high profile name and later in the year he was back, retaining his USBA belt with a twelve round decision over Scott LeDoux.

In March of 1980 Weaver fought John Tate for the WBA title in Tate's backyard of Knoxville, Tennessee. Tate was an amateur star from the 1976 Olympic team, and as a pro had put together a 20-0 record and won the vacant WBA title by decisioning South African Gerrie Coetzee over fifteen rounds in front of 86,000 hostile fans in Pretoria, South

Africa. Tate dominated Weaver for most of fourteen rounds, but with forty seconds left in the fifteenth and final round Weaver caught him with a left hook that dropped Tate to the canvas for the full count.

In October 1980 Weaver made his first defense, traveling to Sun City, South Africa to fight Gerrie Coetzee and knocking him out in the thirteenth round. Coetzee had never previously been down, amateur or pro. The following year he outpointed flashy James "Quick" Tillis over fifteen rounds in Chicago to retain his title after a year's inactivity, and after another year off took on highly regarded Michael Dokes in Las Vegas on December 10, 1982. Dokes came out fast and dropped Weaver inside the opening minute, and as Weaver covered up on the ropes and Dokes missed a few swings referee Joey Curtis stopped the fight and awarded Dokes the title 1:03 into the first round.

In a May 1983 rematch Dokes was awarded a draw after fifteen rounds with Weaver although most, if not all, at ringside felt Weaver had done enough to warrant the verdict comfortably. In June of 1985 Mike took on Pinklon Thomas, who held the WBC title, and lost on an eighth-round knockout. It would be his last title challenge, although a second round knockout of Carl "The Truth" Williams would follow the defeat to Thomas. A win over highly touted South African Johnny DuPlooy was the only real highlight of the later years, and Weaver ended with another knockout defeat to Larry Holmes when he was almost fifty years old.

Mike Weaver now resides in Los Angeles, California and has been working for the U.S. Postal Service since 1999.

CHUCK WEPNER
The Real Rocky

"If I survived the Marines, I can survive Ali."
- *Chuck Wepner*

Chuck Wepner is a former heavyweight boxer from Bayonne, New Jersey who was born on February 26, 1939. As an obscure boxer he went fifteen rounds with world heavyweight champion Muhammad Ali in a 1975 fight, and has often been credited with being the inspiration for Rocky Balboa.

Wepner, nicknamed "The Bayonne Bleeder," debuted as a professional boxer in 1964 and began posting many wins and a few losses. He had formerly boxed while a member of the Marine Corps, and had worked as a security guard before turning pro. He was the New Jersey State Heavyweight Boxing Champion and a popular fighter on the Northeast's Club Boxing circuit. After losing fights to George Foreman (by knockout in three) and Sonny Liston (by knockout in ten) many boxing fans thought his days as a contender were numbered, mainly because after the fight with Liston Wepner needed over 120 sutures in his face. However, after losing to Joe Bugner by a knockout in three in England, Wepner won nine of his next eleven fights including victories over Charlie Polite and former WBA Heavyweight champion Ernie Terrell.

In 1975 it was announced Wepner would challenge Muhammad Ali for the world Heavyweight title. According to a *Time* magazine article called *In Stitches*, Ali was guaranteed $1.5 million and Wepner signed for $100,000. This was considerably more than Wepner had ever earned previously, and therefore he did not need any coaxing. He spent eight weeks training in the Catskill Mountains under the guidance of manager Al Braverman and trainer Bill Prezant, and Prezant prophesied that the fight would be a big surprise. Before the bout a reporter asked Wepner if he thought he could survive in the ring with the champion, to which Wepner replied, "I've been a survivor my whole life... if I survived the Marines, I can survive Ali."

The fight was promoted again by Don King, who promised Ali the then-astonishing sum of $1.5 million. King was by now firmly associated with Ali, and a big player on the heavyweight scene.

In the ninth round Wepner landed a punch to Ali's chest and Ali was knocked down. Wepner went to his corner and said to his manager, "Hey, I knocked him down."

"Yeah," Wepner's manager replied, "but he looks *really* pissed off now..."

In the remaining rounds Ali opened up cuts above both of Wepner's eyes and broke his nose, although the far-behind-in-points Wepner did make a dramatic comeback only to lose in the final minutes.

Young actor Sylvester Stallone watched the fight at home on television and was inspired to write the script for *Rocky* based on Wepner's gutsy challenge. In 1976 Wepner fought professional wrestler André the Giant, which was a scenario similar to the scene in *Rocky III* where Stallone faced Hulk Hogan in the ring.

Chuck Wepner finished his career with a record of thirty-five wins (seventeen by knockout), fourteen losses, and two draws and now lives in Bayonne, New Jersey with his wife Linda.

JO JO WHITE
Down Glory Road

Joseph Henry "Jo Jo" White is a former professional basketball player who was born on November 16, 1946 in St. Louis, Missouri. White, a former Marine private, served as a cornerstone for two Celtics championship teams in the 1970s, making seven All-Star teams before retiring in 1981.

White played college basketball at the University of Kansas, which entered the NCAA Tournament and lost a double overtime thriller to UTEP, then known as Texas Western, in the Midwest regional final (UTEP went on to win the championship, which was depicted in the 2006 Disney film *Glory Road*, with Texas Western playing Kansas in the 1966 college basketball tournament).

After college White played on the 1968 USA Olympic basketball team in Mexico, which went undefeated (9-0) and beat Yugoslavia 65-50 in the title game.

After the Olympics White was drafted in 1969 in the first round (9th pick overall) by the NBA's Boston Celtics who at that time had just won their eleventh championship in thirteen years, but before White even reported to training camp the Celtics' legendary center and player-coach Bill Russell announced his retirement. As a result White would endure a rebuilding season while the Celtics got back on track, drafting Dave Cowens and trading for Paul Silas. Along with these two and veteran John Havlicek, White would be the cornerstone of two Celtic championship teams in the 1970s (1973-74 and 1975-76).

White went on to become one of professional basketball's first "iron men," playing in all eighty-two games for five consecutive seasons during the 1970s, and his skills included great defense, speed, an underrated jump shot, and team leadership.

The 1970 Celtics finished with the franchise's first losing record since 1951, but with White leading the attack from the point guard position the team returned to its winning ways in 1971. He was an All-Star for seven straight years from 1971 through 1977, finishing in the top ten in the league in assists from 1973-77. In 1974 and 1976 White helped lead the Celtics to the NBA championship, and was named the most valuable player of the 1976 NBA Finals. Perhaps the most exciting game White ever played was the triple overtime win against the Phoenix Suns in game five of those finals. He was the game's highest scorer with thirty-three points and had a game high nine assists while leading the Celtics to a 128-126 win. Logging an incredible sixty minutes of playing time, only the Suns' Garfield Heard played more minutes. Many claim it to be the greatest game ever played.

White was traded by the Celtics to the Golden State Warriors in 1979 and retired in 1981 while with the Kansas

City Kings, and on April 9, 1982 his number ten jersey was hung from the rafters at the Boston Garden.

White returned to the Kansas Jayhawks as an assistant coach from 1982-83, continues to be involved in basketball, and is currently director of special projects and community relations with the Celtics while continuing to attend most home games.

TED WILLIAMS
The Splendid Splinter

"The thing I'm most proud of is that I was a Marine Corps fighter pilot." – *Ted Williams*

Theodore Samuel "Ted" Williams (August 30, 1918 - July 5, 2002) was a Major League Baseball player who was born in San Diego, California. He was named after his father, Samuel Stuart Williams, and Teddy Roosevelt. Although at some point the name on his birth certificate was changed to Theodore, his mother and closest friends always called him Teddy. His father was a soldier and sheriff from New York who greatly admired the former president and his mother, May Venzor, was a Salvation Army worker from El Paso, Texas.

Williams, who was variously nicknamed the 'Splendid Splinter,' 'The Kid,' 'Teddy Ballgame,' 'Terrible Ted' and 'The Thumper,' played twenty-one seasons with the Boston

Red Sox although his career was twice interrupted by military service as a Marine Corps pilot.

Williams is widely considered to be one of the greatest hitters in baseball history. He was a two-time American League Most Valuable Player, led the league in batting six times, and won the Triple Crown twice. He had a career batting average of .344, 521 home runs, and was inducted into the Baseball Hall of Fame in 1966. He is the last player in Major League Baseball to bat over .400 in a single season, and holds the highest career batting average of anyone with five hundred or more home runs. His career year was 1941 when he hit .406 with thirty-seven home runs, 120 RBI, and 135 runs scored, and his .551 on base percentage set a record that stood for sixty-one years. An avid sport fisherman as well, he hosted a television show about fishing and was inducted into the IGFA Fishing Hall of Fame.

Williams lived in San Diego's North Park neighborhood and graduated from Herbert Hoover High School in San Diego, where he played baseball. Though he soon had offers from the St. Louis Cardinals and the New York Yankees his mother thought him too young to leave home, so he signed with the local minor league club, the San Diego Padres, while still in high school. He also had a minor league stint with the Minneapolis Millers.

Early in his career Ted stated that he wished to be known as "the greatest hitter who ever lived," an honor he achieved in the eyes of many by the end of his playing days. Williams once stated his goal was to have a father walk down the street with his son, point to him and remark, "Son, there goes the greatest hitter who ever lived." Carl Yastrzemski said of him, "He studied hitting the way a broker studies the stock market."

Williams moved up to the Red Sox in 1939 and made an immediate impact as he led the American League in RBI and finished fourth in MVP balloting. He quickly became known as one of the most potent left-handed hitters in MLB, and a myth soon that developed was that his eyes were the best in history and he was able to read the words on a record album while it was spinning.

In 1941 he entered the last day of the season with a batting average of .39955. This would have been rounded up to .400, making him the first man to hit .400 since Bill Terry in 1930, and manager Joe Cronin left the decision whether to play up to Ted. Williams opted to play in both games of the day's doubleheader and risk losing his record, explaining, "if I can't hit .400 all the way, I don't deserve it." He got six hits in eight at bats to raise his season average to .406, and nobody has done it since.

That season Williams' achievement was overshadowed by Joe DiMaggio's fifty-six-game hitting streak. Their rivalry was played up by the press, but while Williams always felt himself slightly better as a hitter he acknowledged that DiMaggio was the better all-around player.

Another of Williams' memorable accomplishments was his home run off Rip Sewell's notorious 'eephus' pitch during the 1946 All-Star Game at Fenway Park. He challenged Sewell to throw it, and the first one was a strike. Williams challenged Sewell again, and this time hit a home run. In that game he went four for four with two home runs and five RBIs as the AL beat the NL 12-0.

In a climactic ending to his career Ted hit a home run in his very last at bat on September 28, 1960. The classic John Updike essay *Hub Fans Bid Kid Adieu* chronicled this event and is considered one of the greatest pieces of sports writing in American journalism.

Williams served as a pilot during both World War II and the Korean War. He could have received an easy assignment and played baseball for the Navy, but instead joined the V-5 program to become a naval aviator. Williams was first sent to the Navy's Preliminary Ground School at Amherst College for six months of academic instruction in various subjects including math and navigation, where he achieved a 3.85 grade point average. Fellow Red Sox player Johnny Pesky, who went into the same training program, said about him, "Ted mastered intricate problems in fifteen minutes that took the average cadet an hour... and half of the other cadets there were college grads."

Pesky again described Williams' acumen in the advance training for which Pesky personally did not qualify. "I heard Ted literally tore the 'sleeve target' to shreds with his angle dives. He'd shoot from wingovers, zooms and barrel rolls, and after a few passes the sleeve was ribbons. At any rate, I know he broke the all-time record for hits." He then went to Jacksonville for a course in air gunnery, the combat pilot's payoff test, and broke all the records for coordination, reflexes, and visual-reaction time.

Next Williams received preflight training at Athens, Georgia, primary training at NAS Bunker Hill, Indiana, advanced flight training at NAS Pensacola and received his wings and commission in the Marine Corps on May 2, 1944.

Williams first served as a flight instructor at Naval Air Station Pensacola teaching young pilots to fly the F4U Corsair, and was at Pearl Harbor awaiting orders to join the China fleet when the war ended. He finished the war in Hawaii and was released from active duty in January 1946, although he did remain in the reserves.

On May 1, 1952, at the age of thirty-four, Williams was recalled to active duty for service during the Korean War. He hadn't flown for some eight years, but once again turned away offers to sit out the war in comfort as a member of a service baseball team. After eight weeks of refresher flight training and qualification in the F9F Panther jet at Marine Corps Air Station Cherry Point he was assigned to VMF-311, Marine Aircraft Group 33 (MAG-33), which was based at K-3 airfield in Pohang, Korea.

On February 16, 1953 Ted was part of a thirty-five-plane strike against a tank and infantry training school just south of Pyongyang, North Korea. During the mission a piece of flak knocked out his hydraulics and electrical systems, causing him to "limp" his crippled plane back to an Air Force base close to the front lines and land while in flames. For his actions of this day he was awarded the Air Medal.

Williams eventually flew thirty-nine combat missions before being pulled from flight status in June 1953 after the discovery of an inner ear infection which disqualified him from flying. During the war he also served in the same unit as John Glenn, and in the last half of his missions served as Glenn's wingman. While these absences, which took almost five years out of the heart of a great career, significantly limited his career totals, he never publicly complained about the time devoted to military service. Biographer Leigh Montville argues that Williams was not happy about being pressed into service in Korea, but still did what he felt was his patriotic duty.

Williams had a strong respect for General Douglas MacArthur, referring to him as his "idol," and for Williams' fortieth birthday MacArthur sent him an oil painting of himself with the inscription "To Ted Williams - not only America's greatest baseball player, but a great American who served his country. Your friend, Douglas MacArthur. General, U.S. Army."

ARTS
&
ENTERTAINMENT

JAMES BRADY
The Coldest War

"Whatever else we are or may become for the rest of our lives, if you have once been a Marine, you are always a Marine." *– James Brady*

James Winston Brady (November 15, 1928 - January 26, 2009) was an American celebrity columnist who was born in Sheepshead Bay, Brooklyn. He created the *Page Six* gossip column in the *New York Post,* authored the *In Step With* column in *Parade* for nearly twenty-five years, and wrote numerous books about his time in the Marine Corps during the Korean War.

Brady began his career in journalism working as a copy boy for the *Daily News*, where he worked while attending Manhattan College. He left the paper to serve in the Marine Corps during the Korean War, and was a member of the 2nd Battalion, 7th Marines - first leading a rifle platoon, and later acting as the executive officer of a rifle company under future astronaut John Chafee. The majority of his service took place in the North Korean Taebaek Mountains during the fall and bitterly cold winter of 1951-52 during which time he was promoted to First Lieutenant. Brady was awarded the Bronze Star with the Combat 'V' in November of 2001 for his actions on May 31, 1952 in a firefight with Chinese forces near Panmunjom.

Brady wrote extensively about his experiences as a Marine in Korea including his 1990 autobiography *The Coldest War,* which was a Pulitzer Prize finalist. Other books include the 2003 novel *The Marine,* as well as the non-fiction books *The Scariest Place in the World* in 2005 and the 2007 book *Why Marines Fight.* Over the years Brady spoke to many groups of veterans about what he described as a "forgotten war," one where he went to Korea as an immature twenty-three-year-old and "nine months later when I left, I was a grown-up and a pretty good Marine officer."

Clay Felker, the publisher of *New York Magazine*, hired Brady to create its *Intelligencer* column and in 1974 Rupert Murdoch hired him to serve as editor of his new weekly tabloid *Star*, a magazine specializing in celebrity gossip and scandals. Murdoch later shifted Brady to the *New York Post* after he bought the paper in 1976 and he became a major participant in the creation of *Page Six*, a celebrity news and gossip column, by giving the column its name and serving as its first editor.

Starting in 1986 Brady wrote the *In Step With* celebrity profile column in *Parade* and his final column, a profile of actor Kevin Bacon, appeared in the February 15, 2009 issue three weeks after his death. For nearly twenty-five years Brady provided a glimpse into the lives of some of the nation's most beloved celebrities, as well as some up-and-comers who were new to the national spotlight. *Parade* CEO, Chairman, and former Marine Walter Anderson said of him, "Jim was a friend to the seventy-three million Americans who looked forward to his column each week, and was a friend to all of us here at *Parade*. He was also a decorated war hero, and a towering figure in American journalism. He will be extraordinarily missed."

Brady's final book, *Hero of the Pacific: The Life of Legendary Marine John Basilone*, which is about a Marine who was awarded the Medal of Honor winner for his actions at the Battle of Guadalcanal, was completed days before his death and is scheduled to be published in November of 2009.

James Brady died of paralysis on January 26, 2009 at his home in Manhattan at the age of eighty. A cause of death was not immediately disclosed, but he had suffered a stroke several years before. He was survived by his wife of fifty-one years, the former Florence Kelly, two daughters, four grandchildren and a brother.

ART BUCHWALD
Too Soon to Say Goodbye

Arthur Buchwald (October 20, 1925 - January 17, 2007) was an American humorist who was born in Mount Vernon, New York to an Austrian-Hungarian Jewish family, the son of Helen Klineberger and Joseph Buchwald, a curtain manufacturer. He was best known for his long-running political satire and commentary column in *The Washington Post*, which was carried as a syndicated column in many other newspapers.

Buchwald was also known for the *Buchwald v. Paramount* lawsuit, which he and partner Alain Bernheim filed against Paramount Pictures in 1988 in a controversy over the Eddie Murphy film *Coming to America*. Buchwald claimed Paramount had stolen his script treatment, won his case, was awarded damages and accepted a settlement from Paramount.

Buchwald's father put him in the Hebrew Orphan Asylum in New York when the family business failed during the Great Depression and he was moved about between several foster homes, including a Queens boarding house for sick children (he had rickets) which was operated by Seventh-day Adventists. He stayed in the foster home until he was five, at which time he was reunited with his father and sisters and moved to Hollis, a residential community in Queens.

Buchwald did not graduate from Forest Hills High School, and ran away from home at age seventeen. He wanted to join the Marine Corps during World War II but was too young, so he lied about his age and bribed a drunk with half pint of whiskey to sign as his legal guardian. From October of 1942 to October of 1945 he served with the Marines as part of the 4th Marine Aircraft Wing, spent two years in the Pacific Theater, and was discharged as a Sergeant.

On his return Buchwald enrolled at the University of Southern California in Los Angeles on the G.I. Bill despite not having a high school diploma. At USC he was managing editor of the campus magazine *Wampus* and also wrote a column for the college newspaper, the *Daily Trojan*. The university permitted him to continue his studies after learning he had not graduated from high school but deemed him ineligible for a degree, and in an interesting turn of events later awarded him an honorary doctorate in 1993.

In 1948 Buchwald left USC, bought a one-way ticket to Paris, and eventually got a job there as a correspondent for *Variety*. He returned to the United States in 1962 and was syndicated by Tribune Media Services. His column appeared in more than 550 newspapers at its height, he published more than thirty books in his lifetime, and in 1982 his syndicated newspaper column won the Pulitzer Prize for commentary.

Buchwald and his wife Ann, whom he met in Paris, adopted three children and lived in Washington, D.C., although most summers were spent at his house in Vineyard Haven on Martha's Vineyard.

In 2000, at age seventy-four, Buchwald suffered a stroke that left him in the hospital for more than two months, and as a direct result he had a leg amputated below the knee in 2006 due to poor circulation.

In February of 2006 Buchwald checked himself into a Washington, D.C.-area hospice, and although his kidneys were failing he elected to forgo dialysis. In July of that year he returned to his summer home in Tisbury on Martha's Vineyard and while there completed a book about the five months he had spent in the hospice titled *Too Soon to Say Goodbye*. Eulogies which were prepared by his friends, colleagues, and family members which were never delivered (or were not delivered until later) are included in the book.

Art Buchwald died of kidney failure on January 17, 2007 at his son Joel's home in Washington, D.C. at the age of eighty-one. The next day the website of *The New York Times* posted a video obituary in which Buchwald himself declared, "Hi. I'm Art Buchwald, and I just died."

ORVILLE BURRELL
Shaggy

Orville Richard Burrell, better known by his stage name "Shaggy," is a Grammy-nominated reggae singer who was born on October 22, 1968 in Kingston, Jamaica. Burrell takes his nickname from Scooby-Doo's companion. It was given to him by friends during his teenage years because his name bore a similarity to the Scooby Doo character.

His family moved to the United States from Jamaica and settled in the neighborhood of Flatbush in Brooklyn, New York. Shaggy later moved to the town of Valley Stream on Long Island and opened one of his own recording studios there.

In 1988 he joined the Marine Corps and served as a Field Artillery Cannon Crewman with the 5th Battalion, 10th Marines, and while in the Corps served in Operation Desert Storm during the Gulf War. It was during this time that Shaggy perfected his signature singing voice, breaking the constant monotony of running and marching cadences with his flair for inflection. It is also where he got the inspiration for his song *Boombastic*.

Upon his return from the Persian Gulf he decided to pursue a music career, and his first hit came in 1993 with *Oh Carolina*, which was a dancehall re-make of a 'skat' hit by the *Folkes Brothers*. That same year Shaggy appeared on Kenny Dope's hip hop album *The Unreleased Project*. He has worked together with producers such as Sting Intl., Don One (who cut his first track), Lloyd 'Spiderman' Campbell and Robert Livingston. He had further big hits, including *Boombastic* in 1995, which became the theme tune of a popular Levi's commercial.

Shaggy had a major comeback in 2001 featuring worldwide number-one hit singles *It Wasn't Me* and *Angel*, the latter of which was built around two song samples - Merrilee Rush's 1968 hit *Angel of the Morning* (which was remade in 1981 by Juice Newton), and The Steve Miller Band's 1973 hit *The Joker*. The album *Hot Shot*, from which those cuts came, would hit number one on the Billboard 200 and UK album charts.

RICHARD DIEBENKORN
Ocean Park Abstract

Richard Diebenkorn (April 22, 1922 - March 30, 1993) was a well-known painter who was born in Portland, Oregon although his family moved to San Francisco when he was just two years old.

His early work is associated with Abstract Expressionism and the Bay Area Figurative Movement of the 1950s and 1960s. His later work, which is best known as the "Ocean Park paintings," was instrumental to his achievement of worldwide acclaim.

During the late 1940s and early 1950s Diebenkorn lived and worked in various places including New York City, Woodstock, Albuquerque, Urbana, Illinois, and Berkeley, California and developed a style of abstract expressionist painting which captured worldwide attention.

In 1940 Diebenkorn entered Stanford University, and from 1943 to 1945 served in the Marine Corps. After the Second World War the focus of the art world shifted from the School of Paris to the New York School, and in the early 1950s Diebenkorn adopted abstract expressionism as his vehicle for self-expression. He went on to become a leading abstract expressionist on the west coast, and from 1950 to 1952 was enrolled under the G.I. Bill in the University of New Mexico's graduate fine-arts department where he created a lucid version of Abstract Expressionism.

Diebenkorn lived in Berkeley, California from 1955 to 1966. By the mid-1950s he had become an important figurative painter in a style that bridged Henri Matisse with abstract expressionism. Diebenkorn, Elmer Bischoff, Henry Villierme, David Park, James Weeks, and others participated in a renaissance of figurative painting, dubbed the "Bay Area Figurative Movement."

In the fall 1964 through the spring 1965 Diebenkorn traveled through Europe and was granted a cultural visa to visit and view Henri Matisse paintings in important Soviet museums, and when he returned to painting in the Bay Area in mid-1965 his resulting works summed up all that he had learned.

In 1967 Diebenkorn returned to abstraction, this time in a distinctly personal, geometric style that clearly departed from his early abstract expressionist period. The "Ocean Park" series began in 1967 and developed for over twenty-five years, becoming his most famous work and resulting in more than one hundred and forty paintings. Based on the aerial landscape and perhaps the view from the window of his studio, these large-scale abstract compositions are named after the community in Santa Monica, California where he had his studio.

Richard Diebenkorn died due to complications from emphysema in Berkeley on March 30, 1993 at the age of seventy.

ANDRE DUBUS
Giving Up the Gun

Andre Dubus (August 11, 1936 - February 24, 1999) was a short story writer, essayist, and autobiographer who was born in Lake Charles, Louisiana, the oldest child of a Cajun-Irish Catholic family. His surname is pronounced "Duh-BYOOSE," with the accent on the second syllable to rhyme with the noun "excuse." He is recognized as one of the best American short-story writers of the twentieth century.

Dubus grew up in the Bayou country of Lafayette, Louisiana and was educated by the Christian Brothers, a Catholic parochial school which emphasized literature and writing. He graduated from nearby McNeese State College in 1958 as a journalism and English major and then spent six years in the Marine Corps, eventually rising to the rank of captain. After leaving the Marine Corps Dubus moved with his wife and four children to Iowa City, where he eventually graduated from the University of Iowa's Writers' Workshop with an MFA in creative writing.

Although he did write one novel, 1967's *The Lieutenant*, Dubus considered himself to be and is mainly known as a writer of short fiction. Throughout his career he published most of his work in small but distinguished literary journals such as *Ploughshares* and *Sewanee Review*. He was also loyal to a small publishing firm run by David R. Godine that published his first works. When larger book publishers approached him with more financially-rewarding deals, Dubus stayed with Godine. It was only in the last few years of his life, when his medical bills became substantial, that he switched publishers and moved to Alfred A. Knopf.

Dubus' literary career was extensive. His collections include *Separate Flights, Adultery and Other Choices, Finding a Girl in America, The Times Are Never So Bad, Voices from the Moon, The Last Worthless Evening, Selected Stories, Broken Vessels, Dancing After Hours*, and *Meditations from a Movable Chair*. His writing awards include the PEN/Malamud, the Rea Award for the Short Story for excellence in short fiction, the Jean Stein Award from the American Academy of Arts and Letters, and fellowships from the Guggenheim and MacArthur Foundations. Several writing awards are named after Dubus, and his papers are archived at McNeese State University and Xavier University in Louisiana.

Dubus' life was scarred by tragedy. His sister was raped as a young woman, causing him many years of paranoia over his loved ones' safety, and he carried personal firearms to protect himself and those around him until the night in the late 1980s when he almost shot a man in a drunken argument outside a bar in Tuscaloosa, Alabama. In his essay *Giving Up the Gun*, published in *The New Yorker*, Dubus describes that night as the point at which he decided to stop arming himself and to take a less hostile and defensive view of life.

Dubus experienced another personal tragedy late on the night of July 23, 1986 when he was seriously injured in a car accident. He was driving from Boston to his home in Haverhill, Massachusetts when he stopped to assist two disabled motorists, brother and sister Luis and Luz Santiago. As Dubus assisted the injured Luz to the side of the highway an oncoming car swerved and hit them. Luis was killed instantly, and Luz survived only because Dubus had pushed her out of the way. Dubus himself was critically injured when both of his legs were crushed. The left leg had to be amputated above the knee, and Dubus would eventually lose the use of his right leg as well. He would spend three painful years undergoing a series of operations and extensive physical therapy, but despite his efforts to walk with a prosthesis chronic infections confined him to a wheelchair for the remainder of his life. Dubus continued to battle the physical pains imposed by his condition as well as clinical depression, and over the course of his struggles his third wife left him, taking their two young daughters.

To help Dubus with his mounting medical bills his friends and fellow writers, Kurt Vonnegut and John Updike, held a special literary benefit. Dubus was extremely grateful, and his appreciation extended to holding workshops and reading sessions for aspiring writers. Despite these physical, psychological, and emotional difficulties he continued to write, producing two books of essays and a collection of short stories. He also conducted a weekly writers' workshop in his home for a group of young writers, many of whom were teenage girls in a residential program for abused adolescents.

Dubus also found a deeper religious faith at this time. A practicing Catholic all his life, he found that the loss of his mobility drew him closer to God and renewed his Catholic faith at a deeper, personal level. Those who knew him admired the peace and acceptance he had achieved, as well as his ability to live his life without bitterness or self-pity.

Andre Dubus spent his later years in Haverhill until his death from a heart attack in 1999 at the age of sixty-two and is buried in Elmwood Cemetery in Bradford, Massachusetts. He married three times, and fathered a total of six children. His son Andre Dubus III is an author whose most noteworthy book, the novel *House of Sand and Fog*, is a finalist for the National Book Award.

DON & PHIL EVERLY
The Everly Brothers

The Everly Brothers (Don Everly, born Isaac Donald Everly on February 1, 1937 in Brownie, Kentucky and Phil Everly, born Phillip Everly on January 19, 1939 in Chicago, Illinois) are brothers and top-selling country-influenced rock and roll performers known for steel-string guitar playing and close harmony singing. The Everlys are the most successful U.S. rock and roll duo on the Hot One Hundred, with their greatest years coming between 1957 and 1964.

The brothers are both guitarists and use a simple vocal harmony which is mostly based on parallel thirds. With this, each line can often stand on its own as a melody line. This is in contrast to classic harmony lines which, while working well alongside the melody, sound strange by themselves. One example is the song *Devoted to You.*

The duo's harmony singing had a strong influence on rock and roll groups of the 1960s. The Beatles, Beach Boys, and Simon and Garfunkel developed their early singing style by performing Everly covers. In fact the Beatles based the vocal arrangement of *Please Please Me* on *Cathy's Clown.*

Don and Phil's father Ike Everly was a musician and had a show on KMA and KFNF in Shenandoah, Iowa in the 1940s with his wife Margaret and two young sons. Singing on the show gave the brothers their first exposure to the music industry. The family sang together live, and traveled in the area performing as *The Everly Family.*

The Everly Brothers recorded their first single *Keep A' Lovin' Me* in 1956 but it flopped - however their next, *Bye Bye Love*, after being rejected by thirty other acts (including Elvis Presley), reached number two on the pop charts behind Presley's *Let Me Be Your Teddy Bear*, and was number one on the Country and R&B charts. The song became the Everly Brothers' first million-seller. More big hits followed including *Wake Up Little Susie, All I Have to Do Is Dream*, and *Bird Dog.*

The Everly Brothers also toured extensively with Buddy Holly during 1957 and 1958. According to Holly biographer Philip Norman they were responsible for the change in style for Buddy and the Crickets from Levis and t-shirts to the Everly's sharp Ivy League suits. Don also remembers Buddy as a generous songwriter who wrote *Wishing* for them.

Phil Everly was one of Buddy Holly's pallbearers at his funeral in February of 1959. Don Everly did not attend. He later said, "I couldn't go to the funeral. I couldn't go anywhere. I just took to my bed."

Signing with Warner Bros. Records in 1960, the duo continued to have hits. Their first, 1960's *Cathy's Clown* (written by Don and Phil) sold eight million copies and was their biggest-selling record. Shortly after signing with Warner Brothers however, the Everly's fell out with their manager Wesley Rose, who also administered the Acuff-

Rose music publishing company. Consequently, for a period in the early 1960s, the brothers were shut off from Acuff-Rose songwriters. These included Felice and Boudleaux Bryant, who had written the majority of the Everlys' hits, as well as Don and Phil Everly themselves, because they were also contracted to Acuff-Rose as songwriters and had written several of their own hits. With proven sources of hit material unavailable, the Everlys recorded a mix of covers and songs by other writers. Their last U.S. Top Ten hit was 1962's *That's Old Fashioned*, and succeeding years saw the Everly Brothers selling many fewer records in the United States.

Their enlistment in the Marine Corps in November of 1961 also took them out of the spotlight. One of the Everly's few performances during their time in the Corps was an on-leave appearance on *The Ed Sullivan Show* performing *Jezebel* and *Crying In the Rain*. Their star had begun to wane two years before the British Invasion in 1964, although their appeal remained strong in Canada, the United Kingdom, Australia and elsewhere.

After the Marine Corps the brothers resumed their career, but their U.S. chart success was limited. Of the twenty-seven singles the Everly Brothers released on Warner Brothers from 1963 through 1970 only three made the Hot One Hundred, and none peaked higher than thirty-one. They had more success in Britain and Canada however, reaching the top forty in the United Kingdom with singles through 1965, and the top ten in Canada as late as 1967.

Even though the brothers have not produced a studio album since 1989's *Some Hearts* they still tour and perform and have collaborated with other performers, usually singing either backup vocals or duets. Don recorded *Everytime You Leave* with Emmylou Harris in 1979 on her *Blue Kentucky*

Girl album, and in 2006 Phil sang a duet, *Sweet Little Corrina*, with country singer Vince Gill.

Overall the Everly Brothers had twenty-six Billboard Top Forty singles and thirty-five Billboard Top One Hundred singles, hold the record for the most Top One Hundred singles by any duo, and trail only Hall and Oates for the most Top Forty singles by a duo. Hall and Oates had twenty-nine singles in the top forty, all between 1976 and 1990.

In 1986 the Everlys were among the first ten artists inducted into the Rock and Roll Hall of Fame. During the ceremony they were introduced by Neil Young, who observed that every musical group he belonged to had tried and failed to copy the Everly Brothers' harmonies. That year they returned to their boyhood home of Shenandoah to a crowd of 8,500 for a concert, parade, street dedication, class reunion and other activities. Concert fees were donated to The Everly Family Scholarship Fund, which gives scholarships to middle and high school students in Shenandoah every year.

In 1997 the brothers were awarded the Grammy Lifetime Achievement Award. In addition they were inducted into the Country Music Hall of Fame in 2001, the Vocal Group Hall of Fame in 2004, and their pioneering contribution to the genre has been recognized by the Rockabilly Hall of Fame. The Everly Brothers have a star on the Hollywood Walk of Fame at 7000 Hollywood Blvd, and in 2004 *Rolling Stone Magazine* ranked them number thirty-three on their list of the One Hundred Greatest Artists of All Time.

Don and Phil Everly still perform occasionally, despite having declared their retirement.

FREDDY FENDER
Before the Next Teardrop Falls

Freddy Fender (June 4, 1937 - October 14, 2006), born Baldemar Huerta in San Benito, Texas, was a Tejano, country, and rock and roll musician. He performed both as a solo artist and in the groups *Los Super Seven* and the *Texas Tornados,* and is best known for his 1975 hit *Before the Next Teardrop Falls.*

When Fender was a child he and his parents traveled throughout the United States as a circus act. At the age of five he turned a sardine can and some screen door wire into a homemade guitar, and by age ten made his first radio appearance on Harlingen's KGBS-AM radio station where he sang the hit *Paloma Querida* and reportedly won a tub of food worth five dollars.

At the age of sixteen Fender quit school and began a three-year hitch in the Marine Corps. He then returned to Texas and played nightclubs, bars and honky-tonks throughout the south, mostly to Latino audiences, and in 1957 (while known as "El Bebop Kid") he released two songs to moderate success in Mexico and South America - Spanish-language versions of Elvis Presley's *Don't Be Cruel* and Harry Belafonte's *Jamaica Farewell*. He also recorded his own Spanish version of Hank Williams' *Cold Cold Heart* under the title *Tu Frio Corazon*.

Fender became known for his rockabilly music and his cool persona as "Eddie Con Los Shades," and in 1958 changed his name from Baldemar Huerta to Freddy Fender. He took the name Fender from his guitar and amplifier, and Freddy because the alliteration sounded good to him and it would""sell better with Gringos!" He then headed for California.

In 1959 Fender recorded the blues ballad *Wasted Days and Wasted Nights.* The song became popular, but he was beset by legal troubles in May of 1960 after he and a band member were arrested for possession of marijuana in Baton Rouge, Louisiana. After nearly three years in the fearsome Louisiana State Penitentiary Angola prison farm he was released through the intercession of then Governor Jimmie Davis, who was also a songwriter and musician.

By the end of the 1960s Fender was back in Texas working as a mechanic and attending a local junior college, and was only playing music on the weekends until recording the song *Before the Next Teardrop Falls* in 1974. The single was selected for national distribution, and became a number one hit on the Billboard Country and Pop charts. His next three singles, *Secret Love, You'll Lose a Good Thing* and a remake of *Wasted Days and Wasted Nights* all hit the number-one spot on the Billboard Country charts, and between 1975 and 1983 Fender charted a total of twenty-one country hits.

Not only notable for his genre-crossing appeal, more than a few of Fender's hits featured verses or choruses in Spanish.

Rarely did bilingual songs hit the pop charts, and when they did it was more because of a novelty status. Having bilingual songs on the country charts was even more uncommon, given country music's regional insularity and fanbase.

In 1989 Fender teamed up with fellow Tejano/Tex-Mex musicians Doug Sahm, Flaco Jimenez and Augie Meyers to form the Tejano super group the *Texas Tornados.* Their work meshed conjunto, Tejano, R&B, country, and blues to wide acclaim. The group released four albums and won a Grammy in 1990 for 'Best Mexican American Performance' for the track *Soy de San Luis.*

In the late 1990s Fender joined another super group, *Los Super Seven,* with Los Lobos' David Hidalgo and Cesar Rosas, Flaco Jimenez, Ruben Ramos, Joe Ely, and country singer Rick Trevino. In 1998 the group won a Grammy in the Mexican-American Performance category for their self-titled disc.

In 2001 Fender made his final studio recording, a collection of classic Mexican boleros titled *La Música de Baldemar Huerta* that brought him a third Grammy award, this time in the category of Latin Pop Album. Rose Reyes, who worked with Fender in 2004 for a Texas Folklife and Austin tribute titled *Fifty Years of Freddy Fender*, said of the album, "When he did Mexican standards at that point in his career, I expected it to be good because he's a perfectionist. But that record is so beautifully recorded. His voice is perfection. I was so proud it was coming back to his roots."

Fender underwent a kidney transplant in 2002, with the organ being donated by his daughter, and a transplant of the liver in 2004 - but even so his condition continued to worsen. He was suffering from "incurable cancer" tumors on his lungs. On December 31, 2005 Fender performed his last concert and resumed chemotherapy.

Freddy Fender died of lung cancer on October 14, 2006 at his home in Corpus Christi, Texas at the age of sixty-nine with his family at his bedside and is buried in his hometown of San Benito. International news coverage of the death cited an oft-expressed wish by the singer to become the first Mexican-American inducted into the Country Music Hall of Fame, with reporters noting that posthumous induction remains a possibility.

A Freddy Fender Museum and The Conjunto Music Museum opened November 17, 2007 in San Benito. They share a building with The San Benito Historical Museum. His family has committed to continue the Freddy Fender Scholarship Fund and other philanthropic causes about which the musician was passionate.

BILL GALLO
Master Cartoonist

William Gallo is a cartoonist and newspaper columnist for the *New York Daily News* who was born on December 28, 1922 in Manhattan and was the son of a journalist father who died when Bill was just eleven years old.

When Gallo graduated from high school in 1941 he landed a copy boy job on the *Daily News,* and he worked there for seven months until leaving to serve in World War II. Gallo joined the Marine Corps on December 8, 1942 and completed recruit training at Parris Island, South Carolina. He then served in combat in the Pacific Theater with the 20th Marines of the 4th Marine Division and fought at Saipan, Tinian, Kwajalein and Iwo Jima as a demolitions man.

After the war Gallo returned to the *Daily News* and also attended Columbia University and later the School of Visual Arts under the GI Bill of Rights. Then in 1960 he was transferred to the Sports Department of the newspaper, where he began doing sports cartoons... and the rest is history.

Gallo's cartoons were a favorite of sport fans, and he developed the memorable characters "Gunny Sergeant," "Basement Bertha'" and "Yuchie" to illustrate what couldn't be said in a column of text. When asked about them he said, "Basement Bertha started with the Mets in 1962, probably the worst team that ever existed. But they were funny, and they were characters, and they were wonderful, and they were loveable. So I invented this washer-woman type lady who went beyond the unhappiness of losing. She was the happy loser... you know it's not the end of the world for losers as long as you try, and she was always hoping that the Mets would win. And if they lost she'd make a little joke out of it. Yuchie was a kid I knew by the name of Eugene. An Italian kid, and the Italians called Eugene Yuchie, see? This boy was a friend of mine. He was one of the great athletes of the neighborhood, and one of the first ones called to World War II in the Navy. His ship was hit, and he was one of the first casualties. He was killed immediately. I always remembered him and used Yuchie as a remembrance. But the other thing about Yuchie is he is the boy in all of us, growing up in neighborhoods and playing ball. So I remember Yuchie that way. He's the boy in all of us."

JOSH GRACIN
American Idol

Joshua Mario "Josh" Gracin is a country music singer who was born on October 18, 1980 in Westland, Michigan to Mario and Brenda Gracin. While serving in the Marine Corps he gained public attention as the fourth-place finalist on the second season of the Fox Networks talent competition *American Idol.*

After being eliminated from the show Gracin completed his service in the Corps, and after his honorable discharge signed a record deal with Lyric Street Records. His self-titled debut album was released in 2004 and produced a Number One hit, two more Top Five hits on the Billboard Hot Country Songs charts, and was certified gold by the RIAA. His second album, *We Weren't Crazy*, followed in 2008. This album also produced five more chart singles, including a Top Ten for its title track.

Gracin's vocal debut was in an Easter musical presented by his church, and he later appeared in various school productions and talent shows. He then auditioned for and won a national pop orchestra and vocal competition known as the Fairlane Youth Pops Orchestra while a sophomore in high school. During his entire high school career Josh performed at state festivals, fairs and pageants throughout the State of Michigan, and at sixteen he performed on stage at the Grand Ole Opry in a national talent show and recorded a demo CD in Nashville. Upon graduation from John Glenn High School in Westland he attended Western Michigan University before joining the Marine Corps, and after boot camp came home, married, and eventually became a supply clerk at Camp Pendleton near San Diego, California.

Gracin then auditioned for the second season of *American Idol*, a talent competition television program which airs on the Fox Networks. He appeared and competed on the program, finishing fourth place overall, but because of his prior commitment to the Marine Corps did not participate in the subsequent American Idol Finalists concert tour of American venues. Gracin was later sent on a recruiting tour, and made appearances at special events around the United States to promote the Marine Corps. After four years of service he was honorably discharged in September of 2004.

After his *Idol* stint and discharge from the Corps Josh and his wife Anne Marie moved to Tennessee in pursuit of his singing career. They have three daughters and a son.

GEORGE JONES
When Country Wasn't Cool

George Glenn Jones is a country and western singer who was born on September 12, 1931 in Saratoga, Texas and raised in Vidor, Texas along with his brother and five sisters. He is known for his long list of hit records, his distinctive voice and phrasing, and his marriage to Tammy Wynette. Jones was exposed to music from an early age and learned both from his parent's record collection and listening to the gospel music he heard in church. When George was seven the Jones family bought a radio which introduced him to the country music that would become his life. The gift of a guitar when he was a young boy of nine soon had him playing for money on the streets of Beaumont.

Jones left home at sixteen and headed for Jasper, Texas where he found work singing and playing on a local radio station. Before he was out of his teens he married his first wife Dorothy, but their union didn't even last a full year and

George soon joined the Marine Corps. The Korean War being fought at the time, but Jones was not sent overseas. Instead he sang in bars near his base in California, and after leaving the Marine Corps his music career took off.

Jones had a total of fourteen number one country his during his career beginning with 1959's *White Lightning.* Other memorable chart toppers were *She Thinks I Still Care* in 1962, *The Door* in 1975, *(I Was Country) When Country Wasn't Cool* with Barbara Mandrell in 1981, and 1983's *I Always Get Lucky With You.*

Jones alcohol consumption was legendary. For a great part of his life he woke up to a screwdriver and spent the rest of the day drinking bourbon. He was eventually given the nickname "No-Show Jones" as a result of having missed many performances during his days of drug abuse - and the song *No-Show Jones* makes fun of the foibles and weaknesses of Jones and other country singers.

Jones was married twice before he turned twenty-four. His previously mentioned first marriage to Dorothy Bonvillion took place in 1950and lasted but a year, and they had a daughter named Susan. Then in 1954 he married Shirley Ann Corley. This marriage lasted until 1968 and they had two sons, Jeffrey and Brian. He next married fellow country musician Tammy Wynette in 1969. They were married until 1975 and had one daughter, Georgette. Now a published country singer in her own right, she has performed onstage with her famous father.

George Jones married his fourth and current wife, Nancy Sepulveda, on March 4, 1983 in Woodville, Texas and they now live in Franklin, Tennessee.

ROBERT KIYOSAKI
Rich Dad, Poor Dad

Robert Toru Kiyosaki is a well known self-help author, investor, businessman, motivational speaker and inventor who was born on April 8, 1947 in Hawaii. A fourth-generation Japanese American, Kiyosaki is the son of the late educator Ralph H. Kiyosaki.

After graduating from Hilo High School he attended the U.S. Merchant Marine Academy in New York, graduating with the class of 1969 as a deck officer. He later served in the Marine Corps as a helicopter gunship pilot during the Vietnam War and was awarded the Air Medal. Kiyosaki left the Marine Corps in 1974 and got a job selling copy machines for the Xerox Corporation, and then in 1977 he started the company that brought to market the first nylon and Velcro "surfer" wallets. The company was moderately successful at first, but eventually went bankrupt. In the early 1980s Kiyosaki started a business that licensed T-shirts for heavy metal rock bands, and then around 1996 he launched Cashflow Technologies, Inc. which now operates and owns the *Rich Dad* and *Cashflow* brands.

377

Kiyosaki is best known for his *Rich Dad, Poor Dad* series of motivational books and other material, and has written fifteen books with a combined sales volume of over twenty-six million copies. Although he began as a self-publisher the books were subsequently published by Warner Books under the Rich Dad Press imprint. Three of his books, *Rich Dad Poor Dad, Rich Dad's CASHFLOW Quadrant*, and *Rich Dad's Guide to Investing*, have been on the top ten best-seller lists simultaneously in *The Wall Street Journal, USA Today* and the *New York Times*. The book *Rich Kid Smart Kid* was published in 2001 with the intent to help parents teach their children financial concepts.

The central concept of *Rich Dad, Poor Dad* is an anecdotal comparison of his "two fathers." His "poor dad" was his biological father, who became Superintendent of the Hawaii State Department of Education but had very little real net worth. Contrasted with this is his (arguably fictitious) "rich dad," who advocates tax-advantaged investment vehicles such as real estate or small businesses rather than ownership of securities.

A large part of Kiyosaki's teachings focus on generating passive income by means of investment opportunities such as real estate and businesses, with the ultimate goal of being able to support oneself by such investments alone. In tandem with this he defines "assets" as things which generate cash inflow such as rental properties, and "liabilities" as things which *use* cash such as houses, cars, and so on. He also argues that financial leverage is critically important in becoming rich.

Kiyosaki stresses what he calls "financial literacy" as the means to obtaining wealth. He believes life skills are often best learned through experience and that there are important lessons not taught in school. He also says formal education is primarily for those seeking to be employees or self-employed individuals, and that this is an "Industrial Age idea." And according to Kiyosaki, in order to obtain financial freedom one must be either a business owner or an investor and generate passive income.

Robert Kiyosaki is married to Kim Kiyosaki and has one sister, Emi Kiyosaki, who is a Tibetan Buddhist.

ROBERT LUDLUM
The Bourne Identity

Robert Ludlum (May 25, 1927 - March 12, 2001) was a bestselling author who was born in New York City. His father, George Hartford Ludlum, was a businessman who died in 1934 when Robert was just seven years old. Ludlum grew up in New Jersey and was educated privately at the Chesire Academy in Connecticut. Before acting in the comedy *Junior Miss* on Broadway at sixteen Ludlum had already appeared in school theatricals, but his first ambition was to be a football quarterback.

During World War II Ludlum tried to join the Royal Canadian Air Force, but the attempt failed and he instead ended up serving as an infantryman in the Marine Corps from 1945 to 1947. He was posted to the South Pacific during that time, where he wrote a two hundred page manuscript of some of his impressions.

After returning from military service Ludlum attended Wesleyan University and received his B.A. in 1951. In the same year he married actress Mary Ryducha, and they eventually had three children together.

In the 1950s Ludlum worked as a stage and television actor. He was in two hundred television dramas, among them *The Kraft Television Theatre, Studio One*, and *Robert Montgomery Presents,* and was usually cast as a lawyer or a killer. In *The Strong Are Lonely* Ludlum played a soldier, he was Spartacus in *The Gladiator*, and D'Estivel in George Bernard Shaw's *Saint Joan*. Then in 1957 he became a producer at the North Jersey Playhouse in Fort Lee, New Jersey, and in 1960 opened the Playhouse-on-the-Mall in Paramus.

After producing three hundred stage productions for New York and regional theatre Ludlum wrote his first novel, *The Scarlatti Inheritance,* in 1971. It was a tale about Nazis and international financiers and was published after ten rejection slips - but it became an immediate best-seller. The idea for the story came from an old article in the *Illustrated London News* in which one photograph showed a German pushing a wheelbarrow full of inflation banknotes and another picture showed members of the Nazi Party. Ludlum's next thriller, 1973's *The Osterman Weekend*, was later made into a film which was directed by fellow Marine Sam Peckinpah. In the story a television news executive, John Tanner, is recruited by the CIA to reveal a ring of Soviet agents who are perhaps his close friends. Tanner became the prototype of Ludlum's male protagonist - someone who is more lucky and resourceful than the villains ever could guess, and who finds it hard to trust anyone.

From the mid-1970s on Ludlum was a full-time writer. *The Bourne Identity*, released in 1980, started a series of novels in which American counter-assassin Jason Bourne and his nearly superhuman opponent Carlos confront one another in different parts of the world. The character of Carlos was partly based on Venezuelan-born terrorist Ilich Ramirez Sanchez, who in real life was captured in the Sudan in 1994. "Carlos the Jackal" has been linked to the massacre of Israeli athletes at the Munich Olympics in 1972, and other acts of terrorism as well, and is serving a life sentence in a French prison. In *The Bourne Identity* the protagonist is found half-dead and without any memory of who he is. It gradually turns out that he is David Webb, a young Far East scholar who has gotten a new identity from CIA as Jason Bourne to kill Carlos but is betrayed by government officials. Ludlum followed up with the sequels *The Bourne Supremacy* and *The Bourne Ultimatum*. The fourth novel in the series, *The Bourne Legacy,* was written by Eric Van Lustbader three years after Ludlum's death.

In Ludlum's novels multinational right-wing intrigues were often born of economic issues. He also drew parallels between the Nazis and modern day fanatics striving for power. "When the chaos becomes intolerable, it would be their excuse to march in military units and assume the controls, initially with martial law," speculates one of Ludlum's characters in *The Aquitaine Progression*. In *The Matarese Circle* the CIA and KGB join their forces, like the United States and the Soviet Union during World War II, to fight against a circle of terrorists plotting against superpowers. The Matarese dynasty returned again in *The Matarese Countdown* where its members infiltrate the CIA and try to establish a new world economic order.

Ludlum also published books under the pseudonyms Jonathan Ryder (*Trevayne* and *The Cry of the Halidon*) and

Michael Shepherd (*The Road to Gandolpho*) - the latter being written in a humorist style.

The Prometheus Deception, which was released in 2000, was his most prophetic novel. In the story a series of terrorist attacks are used in an international conspiracy to restrict civil rights and increase electronic surveillance for security reasons. The purpose is good - to protect détente and stop wars and crimes. The protagonist is Nicholas Bryson, a deep-cover agent who trusts his instincts while his opponents act mechanically according to their great plan. Bryson has worked years for a shadowy organization called the Directorate. Everybody lies to him, and Ludlum makes it clear to his readers that they should not believe generally accepted truths, world leaders, or the UN Secretary-General. Once again the agent, surrounded by enemies, is fighting himself out of all kinds of corners. He escapes from a ship, a French château full of security men, and a Chinese storehouse. Bryson has many reasons to question the intentions of governmental organizations, the CIA, FBI, and others and shouts in his anger, "The goddamn GRU, the Russians - that's all in the past. Maybe you Cold War cowboys at Langley haven't yet heard the news - the war's over!'

Robert Ludlum died of a heart attack on March 12, 2001 in Naples, Florida at the age of seventy-three.

BUDDY RICH
World's Greatest Drummer

Bernard "Buddy" Rich (September 30, 1917 - April 2, 1987) was a jazz drummer and bandleader who was born in Brooklyn, New York to vaudevillians Robert and Bess Rich. Rich was billed as "the world's greatest drummer," and was known for his virtuosic technique, power and speed.

Arguably the greatest jazz drummer of all time, the legendary Buddy Rich exhibited his love for music through the dedication of his life to the art. His was a career that spanned seven decades, beginning when Rich was eighteen months old and continuing until his death in 1987. Immensely gifted, he could play with remarkable speed and dexterity despite the fact he never received a formal lesson and refused to practice outside of his performances.

Rich was introduced to audiences at a very young age. By 1921, when he was but four, he was a seasoned solo performer with his vaudeville act, *Traps the Drum Wonder*. With his natural sense of rhythm Rich performed regularly on Broadway, and at the peak of his early career was the second-highest paid child entertainer in the world.

Rich's jazz career began in 1937 when he began playing with Joe Marsala at New York's Hickory House. By 1939 he had joined Tommy Dorsey's band, and he later went on to play with such jazz greats as Dizzy Gillespie, Charlie Ventura, Louis Armstrong and Gene Krupa. In 1942, the year after the United States became involved in World War II, Rich left the Dorsey band and volunteered for service in the Marine Corps but was eventually discharged for medical reasons in 1944.

Following his stint in the Marines Rich rejoined Dorsey's band and was featured in several successful Hollywood movies. He was regularly featured in *Jazz at the Philharmonic* during the late 1940s and appeared in such films as *Symphony of Swing* and *How's About It*.

Throughout the 1960s and '70s Rich toured with his own bands and opened two nightclubs, Buddy's Place and Buddy's Place II, and both clubs were regularly filled to capacity by fans of the great master drummer. After opening Buddy's Place II Rich introduced new tunes with elements of rock into his repertoire, demonstrating his ability to adapt to his audience's changing tastes and establishing himself as a great rock drummer as well.

Known for his caustic humor, Rich was a favorite on several television talk shows including the *Tonight Show with Johnny Carson*, the *Mike Douglas Show*, the *Dick Cavett Show* and the *Merv Griffin Show*. During these appearances audiences were entertained by Rich's constant sparring with the hosts and slights of various pop singers.

Buddy Rich received much well deserved recognition

throughout his career, and the *Downbeat Magazine Hall of Fame Award, Modern Drummer Magazine Hall of Fame Award* and *Jazz Unlimited Immortals of Jazz Award* are just a few of his numerous honors. Rich gained international attention for such master compositions as his ten-minute *West Side Story* medley, and during his lengthy career toured around the globe performing for millions of fans and several world leaders including the King of Thailand, King Hussein of Jordan , Queen of England, and U.S. Presidents Franklin Roosevelt, John F. Kennedy and Ronald Reagan.

Buddy Rich died of heart failure on April 2, 1987 following surgery for a malignant brain tumor at the age of sixty-nine and was interred in the Westwood Village Memorial Park Cemetery in Los Angeles, California. Longtime friend Frank Sinatra gave a touching eulogy at his funeral, and today he is remembered as one of history's greatest musicians. According to jazz legend Gene Krupa, Buddy Rich was "the greatest drummer ever to have drawn breath…" and he was.

MARK SHIELDS
On the Campaign Trail

Mark Shields is a well known political columnist and commentator who was born on May 25, 1937 in Weymouth, Massachusetts. Since 1988 Shields has provided weekly political analysis and commentary for PBS' award-winning *The News Hour with Jim Lehrer*. His current sparring partner is David Brooks of *The New York Times*, with previous counterparts being Paul Gigot of the *Wall Street Journal* and syndicated columnist David Gergen. Shields is also a regular panelist on *Inside Washington*, the weekly public affairs show that is seen on both PBS and ABC, and for seventeen years he was moderator and a panelist on CNN's *Capital Gang.*

Shields, a Roman Catholic, graduated from the University of Notre Dame in 1959 and served as an enlisted man in the Marine Corps before coming to Washington in 1965 where he became an aide to Wisconsin Senator William Proxmire. Then in 1968 he went to work for Robert F. Kennedy's presidential campaign, and later held leadership positions in the presidential campaigns of Edmund Muskie and Morris Udall. Over more than a decade he helped manage state and local campaigns in some thirty-eight states, including incumbent Boston Massachusetts Mayor Kevin White's successful re-election campaign in 1975.

In 1979 Shields became an editorial writer for *The Washington Post* and began writing a column the same year which is now distributed nationally by Creators Syndicate. He's covered the last eleven presidential campaigns, and has attended seventeen national party conventions.

A sought after lecturer, Shields has taught American politics at the University of Pennsylvania's Wharton School, Georgetown University's Graduate School of Public Policy, and was a fellow at Harvard's Kennedy Institute of Politics.

Mark Shields is the author of *On the Campaign Trail,* a book about the 1984 presidential campaign.

LEON URIS
Battle Cry

"I have drawn inspiration from the Marine Corps, the Jewish struggle in Palestine and Israel, and the Irish."
–Leon Uris

Leon Uris (August 3, 1924 - June 21, 2003) was a novelist who was born in Baltimore, Maryland, the son of Jewish-American parents Wolf William and Anna (Blumberg) Uris. His father, a Polish-born immigrant, was a paperhanger and then later a storekeeper. William spent a year in Palestine after World War I before entering the United States and derived his surname from "Yerushalmi," meaning "man of Jerusalem." Uris later said of his father, "He was basically a failure. He went from failure to failure."

Leon attended schools in Virginia and Baltimore but never graduated from high school, having failed English three times. At the age of seventeen he joined the Marine Corps and served in the South Pacific as a radioman at Guadalcanal, Tarawa, and New Zealand from 1942 to 1945. While recuperating from malaria in San Francisco he met Betty Beck, a female Marine sergeant, and they married in 1945.

In 1950 *Esquire* magazine bought an article from Uris and this encouraged him to work on a novel. The result was *Battle Cry*, which graphically depicted the toughness and courage of U.S. Marines in the Pacific and was based upon his own personal experiences during the war. He also wrote *The Angry Hills*, a novel set in wartime Greece.

As a screen writer and newspaper correspondent Uris became intensely interested in Israel, which led to his best-known work, *Exodus*, which is about Jewish history from the late nineteenth century through the founding of the state of Israel in 1948. *Exodus* was a worldwide bestseller, was translated into a dozen languages, and was made into a feature film starring Paul Newman in 1960.

Later works include *Mila 18*, a story of the Warsaw ghetto uprising, *Armageddon: A Novel of Berlin*, which reveals the detailed work by British and American intelligence services in planning for the occupation and pacification of post WWII Germany, *Trinity*, an epic novel about Ireland's struggle for independence, *QB VII*, a chilling novel about the role of a Polish doctor in a German concentration camp, and *The Haj*, with insights into the history of the Middle East and the secret machinations of foreigners which have led to today's turmoil.

He also wrote the screenplay for *Battle Cry*, which was made into a motion picture directed by Raoul Walsh and starring James Whitmore, Van Heflin, Aldo Ray, Tab Hunter and Fess Parker. Fittingly the key role of "Mac," the tough as nails platoon sergeant and narrator of the film, went to real life Marine James Whitmore.

Uris was married three times, first to Betty Beck in 1945, with whom he had three children and divorced in 1968, then to Margery Edwards in 1969, who died of an apparent suicide a year later, and finally to Jill Peabody in 1970, with whom he had two children and divorced in 1989.

Leon Uris died of renal failure at his Long Island home on Shelter Island in 2003at the age of seventy-eight.

JOSEPH WAMBAUGH
The Blue Knight

Joseph Aloysius Wambaugh, Jr. is a bestselling author known for his fictional and non-fictional accounts of police work in the United States who was born on January 22, 1937 in East Pittsburgh, Pennsylvania.

The son of a police officer, Wambaugh joined the Marine Corps at age seventeen (an element he works into several of his novels) and served from 1954 to 1958. He received an Associate's degree from Chaffey College, joined the Los Angeles Police Department in 1960 (eventually serving fourteen years) and rose through the ranks from patrolman to detective sergeant. Wambaugh also attended California State University, Los Angeles in his spare time and received Bachelor of Arts and Master of Arts degrees.

Wambaugh's unique perspective on the realities of police work led to his first novel, *The New Centurions*, which was published early in 1971 to critical acclaim and popular success. The success of his early books came while

Wambaugh was still working in the detective division, and he reportedly remarked, "I would have guys in handcuffs asking me for autographs." Soon writing full time, Wambaugh was prolific and popular starting in the 1970s, mixing novels (*The Blue Knight, The Choirboys, The Black Marble*) with nonfiction accounts of crime and detection a.k.a. "true crime" (*The Onion Field*). Later books included *The Delta Star, Lines and Shadows,* and *The Glitter Dome,* which became a TV-movie starring James Garner and John Lithgow.

In contrast to conventionally heroic fictional policemen, Wambaugh brought a gritty texture to his flawed police characters. He changed his approach to his books beginning with *The Choirboys,* employing dark humor and outrageous incidents to emphasize the psychological peril inherent in modern urban police work. Many characters are referenced by often unflattering nicknames rather than given Christian names, while others have almost whimsical names to paint an immediate word portrait for the reader. Wambaugh also became sharply critical of the command structure of the LAPD and some individuals within it, as well as the city government.

Beginning with *The Black Marble* in 1977 Wambaugh devoted at least half of a narrative to satirical, and often biting, observations of the mores and extravagances of the Southern California "rich and famous" lifestyle. *The Black Marble* parodied dog shows and the fading lifestyle of "old" Pasadena, but not entirely unsympathetically. *The Glitter Dome* explored the pornographic film industry, *The Delta Star* delved into the politics and intrigue of the Nobel Prize and scientific research, and *The Secrets of Harry Bright* savaged the Palm Springs lifestyle of wealthy second homes, drugs and drinking, and restricted country clubs.

One of Wambaugh's most famous nonfiction books is *The Blooding*, which tells the story behind how an early landmark case involving DNA fingerprinting helped solve two murders in Leicester, England and resulted in the arrest and conviction of Colin Pitchfork.

In 2003 *Fire Lover: A True Story* brought Wambaugh his second Edgar Award for Best Crime Fact book, and in 2004 he was the recipient of an MWA Grand Master Award. He returned to fiction with *Hollywood Station* in 2006, his first book depicting life in the LAPD since *The Delta Star* in 1983. *Hollywood Station* was highly critical of conditions caused by the federal consent decree under which the LAPD had to operate after the Rampart scandal. In 2008 he followed it with *Hollywood Crows*, featuring many of the same characters in his first sequel to a novel.

Many of his books were made into feature films or TV-movies during the 1970s and '80s. *The Blue Knight*, a novel following the approaching retirement and last working days of aging veteran beat cop "Bumper" Morgan, was made into an Emmy-winning 1973 TV miniseries starring William Holden and later a short-lived TV series starring George Kennedy. His realistic approach to police drama was highly influential in both film and television depictions such as *Hill Street Blues* from the mid-70s onward.

Wambaugh was also involved with creating and developing the NBC series *Police Story*, which ran from 1973 to 1977. The anthology show covered the different aspects of police work (patrol, detective, undercover, etc.) in the LAPD with story ideas and characters supposedly inspired by off-the-record talks with actual police officers. At times the show's characters also dealt with problems not usually seen or associated with typical TV cop shows such as alcohol abuse, adultery and brutality.

Wambaugh was also involved in the production of the acclaimed film versions of *The Onion Field* in 1979 and *The Black Marble* in 1980, with both being directed by Harold Becker. In 1981 he won an Edgar Award from the Mystery Writers of America for his screenplay for the latter film. This was after *The Choirboys* film adaptation had met with very poor critical and audience reception a few years earlier.

Joseph Wambaugh currently resides in Southern California where he recently began teaching screenwriting courses as a guest lecturer for the theater department at the University of California, San Diego.

PHOTO GALLERY

(L) James Gregory as General Ursus in *Planet of the Apes*; (R.) Warren Oates as "Big Toe" Sergeant Hulka in *Stripes*.

ABC's Peter Jennings, NBC's Tom Brokaw and CNN's Bernard Shaw with Walter Cronkite in New York in 1988.

Vice President Al Gore, right, shakes hands with moderator Jim Lehrer, center, as Texas Governor George W. Bush reaches in at the start of the second presidential debate at Wake Forest University in Winston-Salem, NC in October of 11, 2000.

(L) Bill Gallo presents Muhammad Ali with 1974 Fighter of the Year plaque as Joe Louis (R) accepts Man of the Half Century award; (R) Richard Diebenkorn in his studio.

398

(L) Brian Keith as a WWII era Sergeant; (R.) Corporal Drew Carey's ID photo.

Hari Rhodes is "the Man" - and shakes down a pimp in the film *Detroit 9000*.

Ed McMahon wearing khakis (L) and Beatrice Arthur accepting her Emmy (R).

Henry Fonda (L) and Robert Ryan (R) on the set of *Battle of the Bulge*.

(L) George C. Scott as *Patton;* (R) Pernell Roberts (on right) with the cast of *Bonanza.*

(L) R. Lee Ermey as Gunny Hartman in *Full Metal Jacket;* (R) Gene Hackman as Popeye Doyle in *The French Connection.*

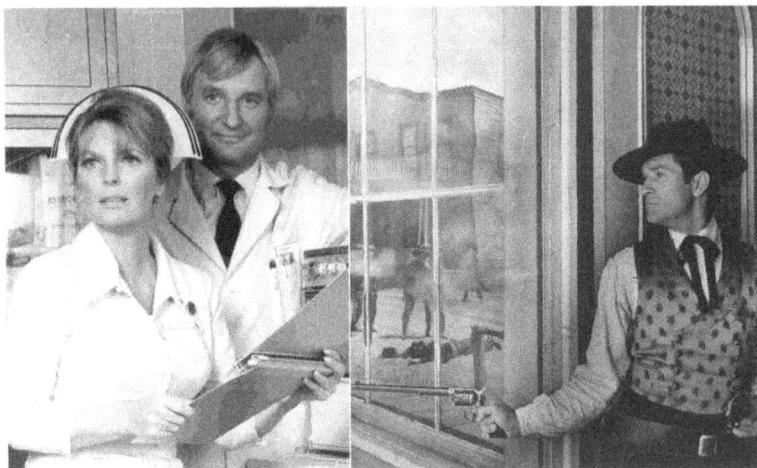

(L) Bobby Troupe and wife Julie London in *Emergency!* (R.) Hugh O'Brian gets the drop on someone in *Wyatt Earp.*

Ben Johnson, Warren Oates, William Holden and Ernest Borgnine in Sam Peckinpah's *The Wild Bunch.*

(Top) Tyrone Power in front of his Curtiss R5C Commando (C-47) in the South Pacific during WWII; (Below) Rob Riggle 'takes five' in Kosovo.

(L) Real life Marine Macdonald Carey plays a Marine in Wake Island; (R) Jonathan Winters performs in a USO show at the Pensacola Civic Center in 1986.

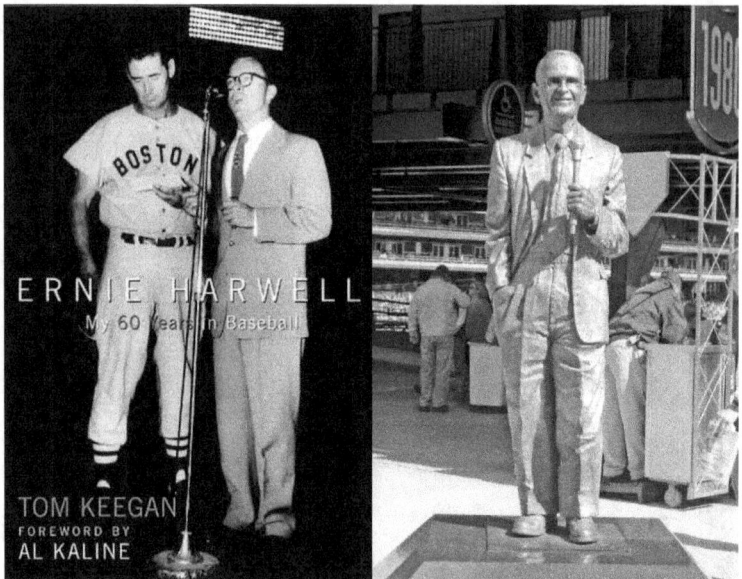

(L) Ernie Harwell chose a photo of himself with fellow Marine Ted Williams for the cover of his book; (R) Statue of Harwell outside Detroit's Comerica Park.

(L) Jerry Coleman suited up and ready to play second base for the New York Yankees; (R) Coleman in the cockpit of his Douglas SBD Dauntless Dive Bomber during WWII.

(L) Rod Carew discusses the art of hitting with Ted Williams; (R) Hank Bauer celebrates another Yankee Championship with teammate Joe DiMaggio.

405

(L) Art Donovan being inducted into the NFL Hall of Fame in Canton, Ohio; (R) Donovan suited up and ready to play for the Baltimore Colts.

(L) Elroy "Crazy Legs" Hirsch shows off his unique style of running; (R) Jo Jo White drives the lane as a member of the Boston Celtics.

Hayden Fry celebrates a Bowl victory with his beloved Iowa Hawkeyes.

Jim Mora and Ron Botchan are being explained the advances in Marine Corps technology by Rich Duff, a National Museum of the Marine Corps docent.

(L) Billy Mills crosses the finish line at the end of the 10,000 meter run during the 1964 Olympic Games; (R) Mills with his Gold medal.

Bob Mathias competing in the high hurdles and discus enroute to a Gold Medal in the Decathlon at the 1948 Olympic Games in London, England.

In Bill Veeck's most famous publicity stunt, 3-foot-7-inch Eddie Gaedel was sent to the plate and earned a walk for the Browns during a regular-season game in 1951.

Lee Trevino had a total of twenty-nine wins on the PGA tour, including six majors.

(L) The Everly Brothers perform on the Ed Sullivan Show in Dress Blues on August 4th, 1957; (R) Singer and *American Idol* contestant Josh Gracin.

Buddy Rich, who was undoubtedly "the greatest drummer who ever lived."

Ted Williams flew a total of thirty-nine combat missions in the Grumman F9F Panther while a member of VMF-311 in Korea.

Boston Red Sox slugger Ted William in cockpit of a Marine F9F-5 Panther jet fighter while taking a refresher course in September of 1952.

New York Mets Hall of Fame pitcher Tom Seaver threw the opening pitch to christen the new Citi Field at the team's inaugural game. Seaver's jersey #41 was retired by the Mets in 1988, and adorns the outfield wall next to Gil Hodges # 14.

(L) Manager Gil Hodges (far left) confers with his starting rotation. Tom Seaver is in the center; (R) Seaver with one of his three Cy Young Awards.

412

Tug McGraw with his son, country music star Tim McGraw.

Rick Monday 'saved the flag' at Dodger Stadium on April 25, 1976. Nice play, Rick!

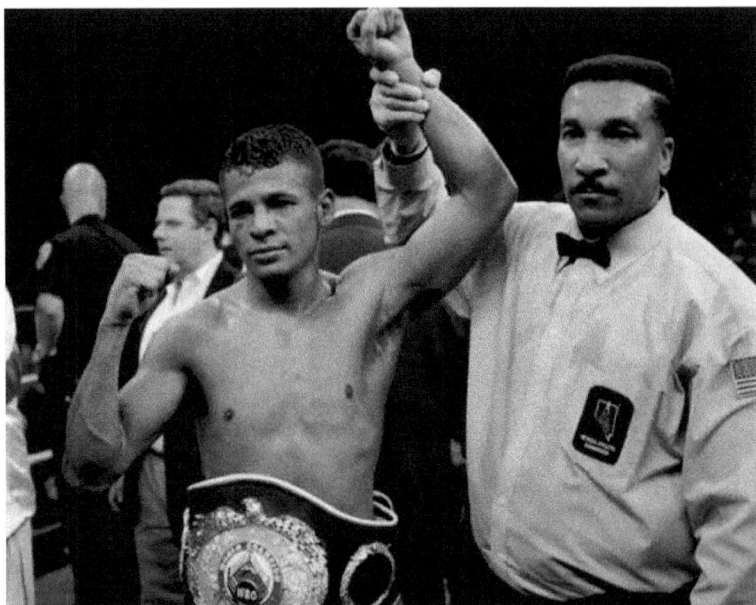

Referee Richard Steele, right, stands with Colombia's Ricardo Torres after Torres won by split decision against Mike Arnaoutis in their WBO world super lightweight title boxing match at the Thomas & Mack Arena in Las Vegas, Saturday, Nov. 18, 2006.

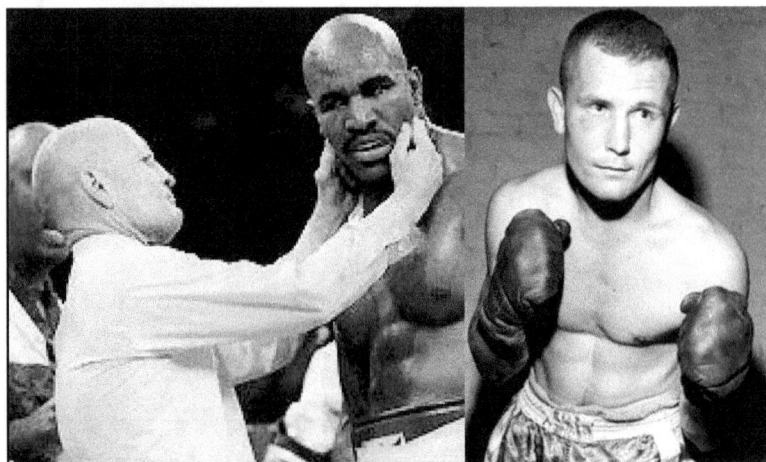

Referee Mills Lane examines Evander Holyfield's ear during his infamous fight with Mike Tyson, and Mills Lane the boxer in his younger years.

(L) Leon Spinks wins the Gold at the 1976 Montreal Olympic Games; (R) Barney Ross, hero of Guadalcanal.

(L) Country music superstar George Jones in the early 1950s; (R) Tommy Loughran strikes a boxing pose.

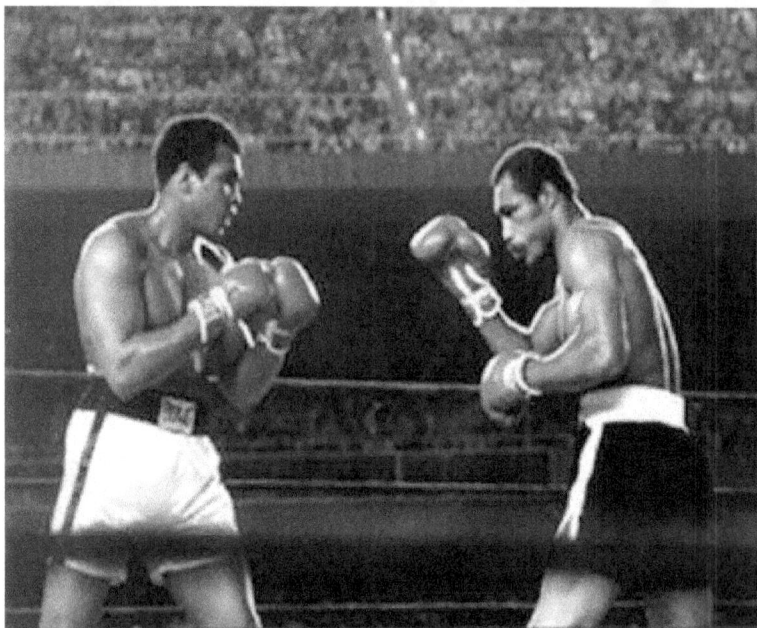

Ken Norton goes toe-to-toe with "The Greatest," Muhammad Ali.

Ken Norton honoring a Marine who received two Purple Hearts in Iraq at World Boxing HOF Tournament Awards Dinner.

(L) Gene Tunney in France during WWI; (R) Tunney in training for a title bout.

(L) The "Real Rocky," Chuck Wepner today; (R) The Bayonne Bleeder in his prime.

Rick Monday with the flag he saved in 1976. Semper Fi!